T0028181

NATIONAL GEOGRAPHIC

BIRDER'S
LIFE LIST
&
JOURNAL

NATIONAL GEOGRAPHIC

WASHINGTON, D.C.

On a list, the Calliope Hummingbird counts the same as any other species, but in life, it is dazzling and unique. A life list keeps track of adventures and memories.

KEEPING A LIFE LIST Noah Strycker

Like a piece of artwork, your life list—a record of the birds you've identified in the wild—starts as an empty canvas. *National Geographic Birder's Life List & Journal* is a blank book, but it is packed with tantalizing possibilities: Of the 830 North American bird species listed here, how many can you see? Some birds, like American Crows and Canada Geese, are common and widespread. Easy-peasy. Others are more specialized. For instance, the skulking Clapper Rail lives in a narrow zone along the Atlantic and Gulf coasts, seldom far from a salt marsh. A few birds require a dedicated mission away from the usual tourist spots. To see an Island Scrub-Jay, you must take a boat to Santa Cruz Island, offshore from Santa Barbara, California, where the jays have been isolated for 150,000 years and have evolved into an endemic species. And if you're keen to add the critically endangered Akikiki to your list, you'll have to travel to Kauai, in the Hawaiian Islands, and carry your binoculars into one of the wettest forests on Earth.

This list derives from the American Birding Association's updated checklist, which covers all 50 United States, Canada, the French islands of Saint Pierre and Miquelon, and adjacent waters up to 200 miles from land. That's a madly diverse swath for any birder to crisscross, from searing deserts to snowcapped mountains, from coral-lined atolls to the Arctic tundra. No two people will end up with the same list of birds, because our lists reflect where we go and who we are. As you use this book and its pages fill up with locations, dates, and notes about sightings, it will tell a story—your story of exploring these realms, one bird at a time.

Strictly speaking, a life list can be kept however you wish. Serious birders adhere to the American Birding Association's *Recording Rules and Interpretations,* a short set of guidelines about what "counts." According to these rules, a bird must be alive, wild, and unrestrained to be listed, so birds seen in zoos are out of bounds. Heard-only birds can go on the list, if a birder desires, as long as they are authentically identified by ear. Birds should be observed respectfully, as described by the ABA's *Code of Birding Ethics.*

No matter who you are or where you live, you can get started right now. Write in the details for each species you've already seen. That's your life list as it stands. The rest of the birds in this book—all the entries left blank for now—are out there at this very moment, occupying their favorite habitats and occasionally wandering to unusual places. Which one will you find next? ■

PAGE 1: In flight, this Rough-legged Hawk has dark "wrists" and a dark belly.
PAGE 3: A male Rose-breasted Grosbeak (top); Yellow-billed Cuckoos (bottom)

Noting the date and location of sightings, especially
of sublime birds like this Great Gray Owl, helps provide
context to what we see—and gives birders a fun mission.

HOW TO USE THIS JOURNAL

National Geographic Birder's Life List & Journal includes every species of bird regularly found in North America north of Mexico, including Hawaii. Here you will find a life list with space for notes, a condensed checklist, a chronological life list, and an index of species and family names. ∎

Life List Pages
Bird species are listed in taxonomic order. Each entry includes a bird's common (English) name, scientific (Latin) name, and unique four-letter code, in addition to a number denoting rarity: 1 for relatively common birds; 2 for range-restricted, secretive, or low-density species; and 3 for rare annual visitors.

Space is provided for you to jot down the date, location, and notes for each sighting. When you see a bird for the first time, write down when and where it happened. It is exciting to expand your life list—and your future self will be grateful to have this information in one place.

Checklist
This section repeats the species list in a condensed format with several blank columns. Here you can track multiple sightings, or divide them into categories. Many birders keep multiple lists: a yard list, a county list, a state list, a birds-seen-while-baking-cookies list … Whatever you fancy, use this checklist to keep track. Or designate each column for a different birding vacation, or a certain year—it's totally flexible.

Chronological Life List
It can be fun to list bird names in order of appearance, recalling your personal birding history by the sequence of species you've spotted. This section provides the space to do so. Add new species here as you identify them.

Index
Birds are indexed by common name. When a bird's name has two or more words, it will be listed by group name with species name following. For instance, Lincoln's Sparrow will be found under S for "Sparrow, Lincoln's." Each species has two page references: the first for the life list page, where you record that species sighting, and the next for the checklist page, where you check it off as you wish. Common names for families also appear in the index.

The Snow Goose, often encountered in blizzard-like
flocks of thousands, has two color morphs.
The dark morph (right of center) or "blue goose"
was considered a separate species until 1983.

DUCKS, GEESE, AND SWANS *(Anatidae)*	When:	Where:
Black-bellied Whistling-Duck *Dendrocygna autumnalis* BBWD 1		

Notes:

	When:	Where:
Fulvous Whistling-Duck *Dendrocygna bicolor* FUWD 1		

Notes:

	When:	Where:
Emperor Goose *Anser canagicus* EMGO 2		

Notes:

	When:	Where:
Snow Goose *Anser caerulescens* SNGO 1		

Notes:

	When:	Where:
Ross's Goose *Anser rossii* ROGO 1		

Notes:

	When:	Where:
Greater White-fronted Goose *Anser albifrons* GWFG 1		
Notes:		

	When:	Where:
Taiga Bean-Goose *Anser fabalis* TABG 3		
Notes:		

	When:	Where:
Tundra Bean-Goose *Anser serrirostris* TUBG 3		
Notes:		

	When:	Where:
Brant *Branta bernicla* BRAN 1		
Notes:		

	When:	Where:
Cackling Goose *Branta hutchinsii* CACG 1		
Notes:		

Canada Goose *Branta canadensis* **CANG 1**	When:	Where:
Notes:		

Hawaiian Goose *Branta sandvicensis* **HAGO 2**	When:	Where:
Notes:		

Mute Swan *Cygnus olor* **MUSW 1**	When:	Where:
Notes:		

Trumpeter Swan *Cygnus buccinator* **TRUS 1**	When:	Where:
Notes:		

Tundra Swan *Cygnus columbianus* **TUSW 1**	When:	Where:
Notes:		

Whooper Swan *Cygnus cygnus* **WHOS 3**	When:	Where:
Notes:		

Egyptian Goose *Alopochen aegyptiaca* **EGGO 2**	When:	Where:
Notes:		

Muscovy Duck *Cairina moschata* **MUDU 2**	When:	Where:
Notes:		

Wood Duck *Aix sponsa* **WODU 1**	When:	Where:
Notes:		

Blue-winged Teal *Spatula discors* **BWTE 1**	When:	Where:
Notes:		

Cinnamon Teal *Spatula cyanoptera* **CITE 1**	When:	Where:
Notes:		

Northern Shoveler *Spatula clypeata* **NSHO 1**	When:	Where:
Notes:		

Gadwall *Mareca strepera* **GADW 1**	When:	Where:
Notes:		

Eurasian Wigeon *Mareca penelope* **EUWI 2**	When:	Where:
Notes:		

American Wigeon *Mareca americana* **AMWI 1**	When:	Where:
Notes:		

	When:	Where:
Laysan Duck *Anas laysanensis* LAYD 3		
Notes:		

	When:	Where:
Hawaiian Duck *Anas wyvilliana* HAWD 2		
Notes:		

	When:	Where:
Mallard *Anas platyrhynchos* MALL 1		
Notes:		

	When:	Where:
Mexican Duck *Anas diazi* MEDU 2		
Notes:		

	When:	Where:
American Black Duck *Anas rubripes* ABDU 1		
Notes:		

Mottled Duck *Anas fulvigula* **MODU 1**	When:	Where:
Notes:		

Northern Pintail *Anas acuta* **NOPI 1**	When:	Where:
Notes:		

Green-winged Teal *Anas crecca* **GWTE 1**	When:	Where:
Notes:		

Canvasback *Aythya valisineria* **CANV 1**	When:	Where:
Notes:		

Redhead *Aythya americana* **REDH 1**	When:	Where:
Notes:		

Common Pochard *Aythya ferina* COMP 3	When:	Where:
Notes:		

Ring-necked Duck *Aythya collaris* RNDU 1	When:	Where:
Notes:		

Tufted Duck *Aythya fuligula* TUDU 3	When:	Where:
Notes:		

Greater Scaup *Aythya marila* GRSC 1	When:	Where:
Notes:		

Lesser Scaup *Aythya affinis* LESC 1	When:	Where:
Notes:		

Steller's Eider *Polysticta stelleri* STEI 2	When:	Where:
Notes:		

Spectacled Eider *Somateria fischeri* SPEI 2	When:	Where:
Notes:		

King Eider *Somateria spectabilis* KIEI 2	When:	Where:
Notes:		

Common Eider *Somateria mollissima* COEI 1	When:	Where:
Notes:		

Harlequin Duck *Histrionicus histrionicus* HADU 1	When:	Where:
Notes:		

Surf Scoter	When:	Where:
Melanitta perspicillata SUSC 1		

Notes:

White-winged Scoter	When:	Where:
Melanitta deglandi WWSC 1		

Notes:

Stejneger's Scoter	When:	Where:
Melanitta stejnegeri STSC 3		

Notes:

Black Scoter	When:	Where:
Melanitta americana BLSC 1		

Notes:

Long-tailed Duck	When:	Where:
Clangula hyemalis LTDU 1		

Notes:

Bufflehead *Bucephala albeola* **BUFF 1**	When:	Where:
Notes:		

Common Goldeneye *Bucephala clangula* **COGO 1**	When:	Where:
Notes:		

Barrow's Goldeneye *Bucephala islandica* **BAGO 1**	When:	Where:
Notes:		

Smew *Mergellus albellus* **SMEW 3**	When:	Where:
Notes:		

Hooded Merganser *Lophodytes cucullatus* **HOME 1**	When:	Where:
Notes:		

Common Merganser *Mergus merganser* **COME 1**	When:	Where:
Notes:		

Red-breasted Merganser *Mergus serrator* **RBME 1**	When:	Where:
Notes:		

Masked Duck *Nomonyx dominicus* **MADU 3**	When:	Where:
Notes:		

Ruddy Duck *Oxyura jamaicensis* **RUDU 1**	When:	Where:
Notes:		

CURASSOWS AND GUANS *(Cracidae)* **Plain Chachalaca** *Ortalis vetula* **PLCH 2**	When:	Where:
Notes:		

NEW WORLD QUAIL (Odontophoridae)	When:	Where:

Mountain Quail
Oreortyx pictus   MOUQ 1

Notes:

	When:	Where:

Northern Bobwhite
Colinus virginianus   NOBO 1

Notes:

	When:	Where:

Scaled Quail
Callipepla squamata   SCQU 1

Notes:

	When:	Where:

California Quail
Callipepla californica   CAQU 1

Notes:

	When:	Where:

Gambel's Quail
Callipepla gambelii   GAQU 1

Notes:

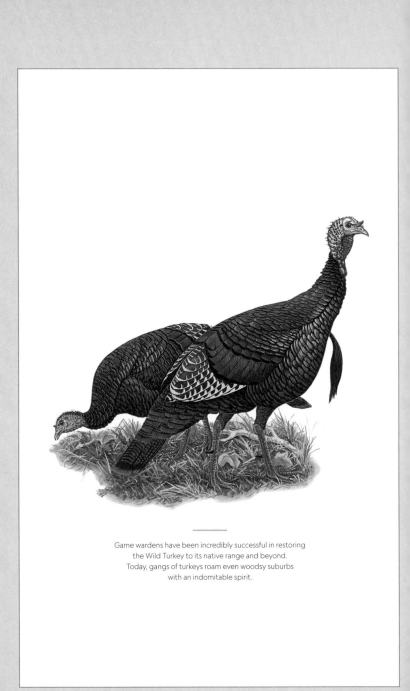

Game wardens have been incredibly successful in restoring
the Wild Turkey to its native range and beyond.
Today, gangs of turkeys roam even woodsy suburbs
with an indomitable spirit.

Montezuma Quail
Cyrtonyx montezumae   MONQ 2

When:

Where:

Notes:

PARTRIDGES, GROUSE, TURKEYS, AND OLD WORLD QUAIL *(Phasianidae)*

When:

Where:

Wild Turkey
Meleagris gallopavo   WITU 1

Notes:

Ruffed Grouse
Bonasa umbellus   RUGR 1

When:

Where:

Notes:

Spruce Grouse
Canachites canadensis   SPGR 2

When:

Where:

Notes:

Willow Ptarmigan
Lagopus lagopus   WIPT 2

When:

Where:

Notes:

Rock Ptarmigan *Lagopus muta* **ROPT 2**	When:	Where:
Notes:		

White-tailed Ptarmigan *Lagopus leucura* **WTPT 2**	When:	Where:
Notes:		

Greater Sage-Grouse *Centrocercus urophasianus* **GRSG 1**	When:	Where:
Notes:		

Gunnison Sage-Grouse *Centrocercus minimus* **GUSG 2**	When:	Where:
Notes:		

Dusky Grouse *Dendragapus obscurus* **DUGR 2**	When:	Where:
Notes:		

Sooty Grouse *Dendragapus fuliginosus* **SOGR 2**	When:	Where:
Notes:		

Sharp-tailed Grouse *Tympanuchus phasianellus* **STGR 2**	When:	Where:
Notes:		

Greater Prairie-Chicken *Tympanuchus cupido* **GRPC 2**	When:	Where:
Notes:		

Lesser Prairie-Chicken *Tympanuchus pallidicinctus* **LEPC 2**	When:	Where:
Notes:		

Gray Partridge *Perdix perdix* **GRAP 2**	When:	Where:
Notes:		

Ring-necked Pheasant *Phasianus colchicus* **RNEP 1**	When:	Where:
Notes:		

Kalij Pheasant *Lophura leucomelanos* **KAPH 2**	When:	Where:
Notes:		

Indian Peafowl *Pavo cristatus* **INPE 2**	When:	Where:
Notes:		

Gray Francolin *Francolinus pondicerianus* **GRAF 2**	When:	Where:
Notes:		

Black Francolin *Francolinus francolinus* **BLFR 2**	When:	Where:
Notes:		

Red Junglefowl
Gallus gallus   **REJU 2**

When: Where:

Notes:

Himalayan Snowcock
Tetraogallus himalayensis   **HISN 2**

When: Where:

Notes:

Chukar
Alectoris chukar   **CHUK 2**

When: Where:

Notes:

Erckel's Francolin
Pternistis erckelii   **ERFR 2**

When: Where:

Notes:

FLAMINGOS *(Phoenicopteridae)*

American Flamingo
Phoenicopterus ruber   **AMFL 3**

When: Where:

Notes:

GREBES *(Podicipedidae)*	When:	Where:
Least Grebe *Tachybaptus dominicus* **LEGR 2**		

Notes:

Pied-billed Grebe *Podilymbus podiceps* **PBGR 1**	When:	Where:

Notes:

Horned Grebe *Podiceps auritus* **HOGR 1**	When:	Where:

Notes:

Red-necked Grebe *Podiceps grisegena* **RNGR 1**	When:	Where:

Notes:

Eared Grebe *Podiceps nigricollis* **EAGR 1**	When:	Where:

Notes:

Western Grebe *Aechmophorus occidentalis* **WEGR 1**	When:	Where:

Notes:

Clark's Grebe *Aechmophorus clarkii* **CLGR 1**	When:	Where:

Notes:

SANDGROUSES (*Pteroclidae*) **Chestnut-bellied Sandgrouse** *Pterocles exustus* **CBSA 3**	When:	Where:

Notes:

PIGEONS AND DOVES (*Columbidae*) **Rock Pigeon** *Columba livia* **ROPI 1**	When:	Where:

Notes:

White-crowned Pigeon *Patagioenas leucocephala* **WCPI 2**	When:	Where:

Notes:

	When:	Where:
Red-billed Pigeon *Patagioenas flavirostris* **RBPI 2**		

Notes:

	When:	Where:
Band-tailed Pigeon *Patagioenas fasciata* **BTPI 1**		

Notes:

	When:	Where:
Eurasian Collared-Dove *Streptopelia decaocto* **EUCD 1**		

Notes:

	When:	Where:
Spotted Dove *Streptopelia chinensis* **SPDO 2**		

Notes:

	When:	Where:
Zebra Dove *Geopelia striata* **ZEBD 2**		

Notes:

Inca Dove	When:	Where:
Columbina inca **INDO 1**		

Notes:

Common Ground Dove	When:	Where:
Columbina passerina **CGDO 1**		

Notes:

Ruddy Ground Dove	When:	Where:
Columbina talpacoti **RGDO 3**		

Notes:

White-tipped Dove	When:	Where:
Leptotila verreauxi **WTDO 2**		

Notes:

White-winged Dove	When:	Where:
Zenaida asiatica **WWDO 1**		

Notes:

Mourning Dove *Zenaida macroura* MODO 1	When:	Where:
Notes:		

CUCKOOS, ROADRUNNERS, AND ANIS *(Cuculidae)* **Smooth-billed Ani** *Crotophaga ani* SBAN 3	When:	Where:
Notes:		

Groove-billed Ani *Crotophaga sulcirostris* GBAN 2	When:	Where:
Notes:		

Greater Roadrunner *Geococcyx californianus* GRRO 1	When:	Where:
Notes:		

Common Cuckoo *Cuculus canorus* COCU 3	When:	Where:
Notes:		

Yellow-billed Cuckoo *Coccyzus americanus* **YBCU 1**	When:	Where:

Notes:

Mangrove Cuckoo *Coccyzus minor* **MACU 2**	When:	Where:

Notes:

Black-billed Cuckoo *Coccyzus erythropthalmus* **BBCU 1**	When:	Where:

Notes:

GOATSUCKERS *(Caprimulgidae)*

Lesser Nighthawk *Chordeiles acutipennis* **LENI 1**	When:	Where:

Notes:

Common Nighthawk *Chordeiles minor* **CONI 1**	When:	Where:

Notes:

| **Antillean Nighthawk** | When: | Where: |
| *Chordeiles gundlachii*   ANNI 2 | | |

Notes:

| **Common Pauraque** | When: | Where: |
| *Nyctidromus albicollis*   COPA 2 | | |

Notes:

| **Common Poorwill** | When: | Where: |
| *Phalaenoptilus nuttallii*   COPO 1 | | |

Notes:

| **Chuck-will's-widow** | When: | Where: |
| *Antrostomus carolinensis*   CWWI 1 | | |

Notes:

| **Buff-collared Nightjar** | When: | Where: |
| *Antrostomus ridgwayi*   BCNI 3 | | |

Notes:

On hot summer nights, the nocturnal Chuck-will's-widow
hunts on the wing for insects—and occasionally small bats
and songbirds—throughout the American Southeast.
Like that of its smaller relative, the Eastern Whip-poor-will,
its name is onomatopoeic.

	When:	Where:
Eastern Whip-poor-will *Antrostomus vociferus*　**EWPW 1**		
Notes:		

	When:	Where:
Mexican Whip-poor-will *Antrostomus arizonae*　**MWPW 2**		
Notes:		

	When:	Where:
SWIFTS *(Apodidae)* **Black Swift** *Cypseloides niger*　**BLSW 2**		
Notes:		

	When:	Where:
Chimney Swift *Chaetura pelagica*　**CHSW 1**		
Notes:		

	When:	Where:
Vaux's Swift *Chaetura vauxi*　**VASW 1**		
Notes:		

Mariana Swiftlet *Aerodramus bartschi* **MASW 2**	When:	Where:
Notes:		

White-throated Swift *Aeronautes saxatalis* **WTSW 1**	When:	Where:
Notes:		

HUMMINGBIRDS *(Trochilidae)* **Mexican Violetear** *Colibri thalassinus* **MEVI 3**	When:	Where:
Notes:		

Rivoli's Hummingbird *Eugenes fulgens* **RIHU 2**	When:	Where:
Notes:		

Blue-throated Mountain-gem *Lampornis clemenciae* **BTMG 2**	When:	Where:
Notes:		

| **Lucifer Hummingbird**
Calothorax lucifer   **LUHU 2** | When: | Where: |
| Notes: | | |

| **Ruby-throated Hummingbird**
Archilochus colubris   **RTHU 1** | When: | Where: |
| Notes: | | |

| **Black-chinned Hummingbird**
Archilochus alexandri   **BCHU 1** | When: | Where: |
| Notes: | | |

| **Anna's Hummingbird**
Calypte anna   **ANHU 1** | When: | Where: |
| Notes: | | |

| **Costa's Hummingbird**
Calypte costae   **COHU 1** | When: | Where: |
| Notes: | | |

Calliope Hummingbird *Selasphorus calliope* **CAHU 1**	When:	Where:
Notes:		

Rufous Hummingbird *Selasphorus rufus* **RUHU 1**	When:	Where:
Notes:		

Allen's Hummingbird *Selasphorus sasin* **ALHU 1**	When:	Where:
Notes:		

Broad-tailed Hummingbird *Selasphorus platycercus* **BTHU 1**	When:	Where:
Notes:		

Broad-billed Hummingbird *Cynanthus latirostris* **BBIH 2**	When:	Where:
Notes:		

White-eared Hummingbird *Basilinna leucotis* **WEHU 3**	When:	Where:
Notes:		

Violet-crowned Hummingbird *Leucolia violiceps* **VCHU 2**	When:	Where:
Notes:		

Berylline Hummingbird *Saucerottia beryllina* **BEHU 3**	When:	Where:
Notes:		

Buff-bellied Hummingbird *Amazilia yucatanensis* **BBEH 2**	When:	Where:
Notes:		

RAILS, GALLINULES, AND COOTS (Rallidae) **Ridgway's Rail** *Rallus obsoletus* **RIRA 2**	When:	Where:
Notes:		

Clapper Rail *Rallus crepitans* **CLRA 1**	When:	Where:
Notes:		

King Rail *Rallus elegans* **KIRA 1**	When:	Where:
Notes:		

Virginia Rail *Rallus limicola* **VIRA 1**	When:	Where:
Notes:		

Sora *Porzana carolina* **SORA 1**	When:	Where:
Notes:		

Common Gallinule *Gallinula galeata* **COGA 1**	When:	Where:
Notes:		

	When:	Where:
Hawaiian Coot *Fulica alai* **HACO 2**		
Notes:		

	When:	Where:
American Coot *Fulica americana* **AMCO 1**		
Notes:		

	When:	Where:
Purple Gallinule *Porphyrio martinicus* **PUGA 1**		
Notes:		

	When:	Where:
Purple Swamphen *Porphyrio porphyrio* **PUSW 2**		
Notes:		

	When:	Where:
Yellow Rail *Coturnicops noveboracensis* **YERA 2**		
Notes:		

Black Rail *Laterallus jamaicensis* **BLRA 2**	When:	Where:
Notes:		

LIMPKINS (*Aramidae*) **Limpkin** *Aramus guarauna* **LIMP 2**	When:	Where:
Notes:		

CRANES (*Gruidae*) **Sandhill Crane** *Antigone canadensis* **SACR 1**	When:	Where:
Notes:		

Whooping Crane *Grus americana* **WHCR 2**	When:	Where:
Notes:		

STILTS AND AVOCETS (*Recurvirostridae*) **Black-necked Stilt** *Himantopus mexicanus* **BNST 1**	When:	Where:
Notes:		

| **American Avocet** | When: | Where: |
| *Recurvirostra americana*   **AMAV 1** | | |

Notes:

OYSTERCATCHERS (*Haematopodidae*)	When:	Where:
American Oystercatcher		
Haematopus palliatus **AMOY 1**		

Notes:

| **Black Oystercatcher** | When: | Where: |
| *Haematopus bachmani*   **BLOY 1** | | |

Notes:

LAPWINGS AND PLOVERS (*Charadriidae*)	When:	Where:
Black-bellied Plover		
Pluvialis squatarola **BBPL 1**		

Notes:

| **American Golden-Plover** | When: | Where: |
| *Pluvialis dominica*   **AMGP 1** | | |

Notes:

| **Pacific Golden-Plover** | When: | Where: |
| *Pluvialis fulva* **PAGP 2** | | |

Notes:

| **Killdeer** | When: | Where: |
| *Charadrius vociferus* **KILL 1** | | |

Notes:

| **Common Ringed Plover** | When: | Where: |
| *Charadrius hiaticula* **CRPL 3** | | |

Notes:

| **Semipalmated Plover** | When: | Where: |
| *Charadrius semipalmatus* **SEPL 1** | | |

Notes:

| **Piping Plover** | When: | Where: |
| *Charadrius melodus* **PIPL 2** | | |

Notes:

	When:	Where:
Lesser Sand-Plover *Charadrius mongolus* **LSAP 3**		

Notes:

	When:	Where:
Wilson's Plover *Charadrius wilsonia* **WIPL 1**		

Notes:

	When:	Where:
Mountain Plover *Charadrius montanus* **MOPL 2**		

Notes:

	When:	Where:
Snowy Plover *Charadrius nivosus* **SNPL 1**		

Notes:

SANDPIPERS, PHALAROPES, AND ALLIES *(Scolopacidae)*	When:	Where:
Upland Sandpiper *Bartramia longicauda* **UPSA 1**		

Notes:

Bristle-thighed Curlew *Numenius tahitiensis* **BTCU 2**	When:	Where:
Notes:		

Whimbrel *Numenius phaeopus* **WHIM 1**	When:	Where:
Notes:		

Long-billed Curlew *Numenius americanus* **LBCU 1**	When:	Where:
Notes:		

Bar-tailed Godwit *Limosa lapponica* **BTGO 2**	When:	Where:
Notes:		

Black-tailed Godwit *Limosa limosa* **BLTG 3**	When:	Where:
Notes:		

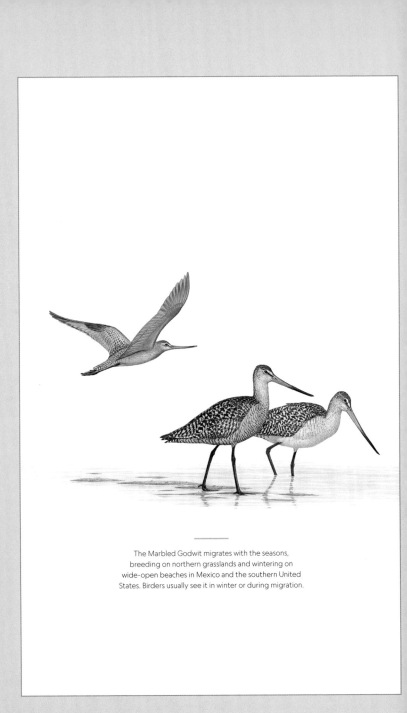

The Marbled Godwit migrates with the seasons,
breeding on northern grasslands and wintering on
wide-open beaches in Mexico and the southern United
States. Birders usually see it in winter or during migration.

Hudsonian Godwit *Limosa haemastica* **HUGO 1**	When:	Where:
Notes:		

Marbled Godwit *Limosa fedoa* **MAGO 1**	When:	Where:
Notes:		

Ruddy Turnstone *Arenaria interpres* **RUTU 1**	When:	Where:
Notes:		

Black Turnstone *Arenaria melanocephala* **BLTU 1**	When:	Where:
Notes:		

Red Knot *Calidris canutus* **REKN 1**	When:	Where:
Notes:		

| **Surfbird** | When: | Where: |
| *Calidris virgata*   **SURF 1** | | |

Notes:

| **Ruff** | When: | Where: |
| *Calidris pugnax*   **RUFF 3** | | |

Notes:

| **Sharp-tailed Sandpiper** | When: | Where: |
| *Calidris acuminata*   **SPTS 3** | | |

Notes:

| **Stilt Sandpiper** | When: | Where: |
| *Calidris himantopus*   **STSA 1** | | |

Notes:

| **Curlew Sandpiper** | When: | Where: |
| *Calidris ferruginea*   **CUSA 3** | | |

Notes:

Temminck's Stint *Calidris temminckii* TEST 3	When:	Where:
Notes:		

Long-toed Stint *Calidris subminuta* LTST 3	When:	Where:
Notes:		

Red-necked Stint *Calidris ruficollis* RNST 3	When:	Where:
Notes:		

Sanderling *Calidris alba* SAND 1	When:	Where:
Notes:		

Dunlin *Calidris alpina* DUNL 1	When:	Where:
Notes:		

Rock Sandpiper *Calidris ptilocnemis* ROSA 2	When:	Where:
Notes:		

Purple Sandpiper *Calidris maritima* PUSA 1	When:	Where:
Notes:		

Baird's Sandpiper *Calidris bairdii* BASA 1	When:	Where:
Notes:		

Least Sandpiper *Calidris minutilla* LESA 1	When:	Where:
Notes:		

White-rumped Sandpiper *Calidris fuscicollis* WRSA 1	When:	Where:
Notes:		

Buff-breasted Sandpiper *Calidris subruficollis* **BBSA 1**	When:	Where:
Notes:		

Pectoral Sandpiper *Calidris melanotos* **PESA 1**	When:	Where:
Notes:		

Semipalmated Sandpiper *Calidris pusilla* **SESA 1**	When:	Where:
Notes:		

Western Sandpiper *Calidris mauri* **WESA 1**	When:	Where:
Notes:		

Short-billed Dowitcher *Limnodromus griseus* **SBDO 1**	When:	Where:
Notes:		

Long-billed Dowitcher *Limnodromus scolopaceus* **LBDO 1**	When:	Where:
Notes:		

American Woodcock *Scolopax minor* **AMWO 1**	When:	Where:
Notes:		

Common Snipe *Gallinago gallinago* **COSN 3**	When:	Where:
Notes:		

Wilson's Snipe *Gallinago delicata* **WISN 1**	When:	Where:
Notes:		

Terek Sandpiper *Xenus cinereus* **TESA 3**	When:	Where:
Notes:		

Common Sandpiper *Actitis hypoleucos* COSA 3	When:	Where:
Notes:		

Spotted Sandpiper *Actitis macularius* SPSA 1	When:	Where:
Notes:		

Solitary Sandpiper *Tringa solitaria* SOSA 1	When:	Where:
Notes:		

Gray-tailed Tattler *Tringa brevipes* GTTA 3	When:	Where:
Notes:		

Wandering Tattler *Tringa incana* WATA 1	When:	Where:
Notes:		

Lesser Yellowlegs *Tringa flavipes* **LEYE 1**	When:	Where:
Notes:		

Willet *Tringa semipalmata* **WILL 1**	When:	Where:
Notes:		

Common Greenshank *Tringa nebularia* **COMG 3**	When:	Where:
Notes:		

Greater Yellowlegs *Tringa melanoleuca* **GRYE 1**	When:	Where:
Notes:		

Wood Sandpiper *Tringa glareola* **WOSA 2**	When:	Where:
Notes:		

Wilson's Phalarope *Phalaropus tricolor* **WIPH 1**	When:	Where:
Notes:		

Red-necked Phalarope *Phalaropus lobatus* **RNPH 1**	When:	Where:
Notes:		

Red Phalarope *Phalaropus fulicarius* **REPH 1**	When:	Where:
Notes:		

SKUAS AND JAEGERS *(Stercorariidae)* **Great Skua** *Stercorarius skua* **GRSK 3**	When:	Where:
Notes:		

South Polar Skua *Stercorarius maccormicki* **SPSK 2**	When:	Where:
Notes:		

Pomarine Jaeger *Stercorarius pomarinus* POJA 1	When:	Where:
Notes:		

Parasitic Jaeger *Stercorarius parasiticus* PAJA 1	When:	Where:
Notes:		

Long-tailed Jaeger *Stercorarius longicaudus* LTJA 1	When:	Where:
Notes:		

AUKS, MURRES, AND PUFFINS *(Alcidae)* **Dovekie** *Alle alle* DOVE 2	When:	Where:
Notes:		

Common Murre *Uria aalge* COMU 1	When:	Where:
Notes:		

Thick-billed Murre *Uria lomvia* **TBMU 1**	When:	Where:
Notes:		

Razorbill *Alca torda* **RAZO 1**	When:	Where:
Notes:		

Black Guillemot *Cepphus grylle* **BLGU 1**	When:	Where:
Notes:		

Pigeon Guillemot *Cepphus columba* **PIGU 1**	When:	Where:
Notes:		

Long-billed Murrelet *Brachyramphus perdix* **LBMU 3**	When:	Where:
Notes:		

Marbled Murrelet _Brachyramphus marmoratus_ **MAMU 1**	When:	Where:
Notes:		

Kittlitz's Murrelet _Brachyramphus brevirostris_ **KIMU 2**	When:	Where:
Notes:		

Scripps's Murrelet _Synthliboramphus scrippsi_ **SCMU 2**	When:	Where:
Notes:		

Guadalupe Murrelet _Synthliboramphus hypoleucus_ **GUMU 3**	When:	Where:
Notes:		

Craveri's Murrelet _Synthliboramphus craveri_ **CRMU 3**	When:	Where:
Notes:		

Ancient Murrelet	When:	Where:
Synthliboramphus antiquus **ANMU 2**		

Notes:

Cassin's Auklet	When:	Where:
Ptychoramphus aleuticus **CAAU 1**		

Notes:

Parakeet Auklet	When:	Where:
Aethia psittacula **PAAU 2**		

Notes:

Least Auklet	When:	Where:
Aethia pusilla **LEAU 2**		

Notes:

Whiskered Auklet	When:	Where:
Aethia pygmaea **WHAU 2**		

Notes:

Crested Auklet *Aethia cristatella* **CRAU 2**	When:	Where:
Notes:		

Rhinoceros Auklet *Cerorhinca monocerata* **RHAU 1**	When:	Where:
Notes:		

Atlantic Puffin *Fratercula arctica* **ATPU 1**	When:	Where:
Notes:		

Horned Puffin *Fratercula corniculata* **HOPU 1**	When:	Where:
Notes:		

Tufted Puffin *Fratercula cirrhata* **TUPU 1**	When:	Where:
Notes:		

GULLS, TERNS, AND SKIMMERS *(Laridae)*	When:	Where:
Black-legged Kittiwake *Rissa tridactyla* **BLKI 1**		

Notes:

	When:	Where:
Red-legged Kittiwake *Rissa brevirostris* **RLKI 2**		

Notes:

	When:	Where:
Ivory Gull *Pagophila eburnea* **IVGU 3**		

Notes:

	When:	Where:
Sabine's Gull *Xema sabini* **SAGU 1**		

Notes:

	When:	Where:
Bonaparte's Gull *Chroicocephalus philadelphia* **BOGU 1**		

Notes:

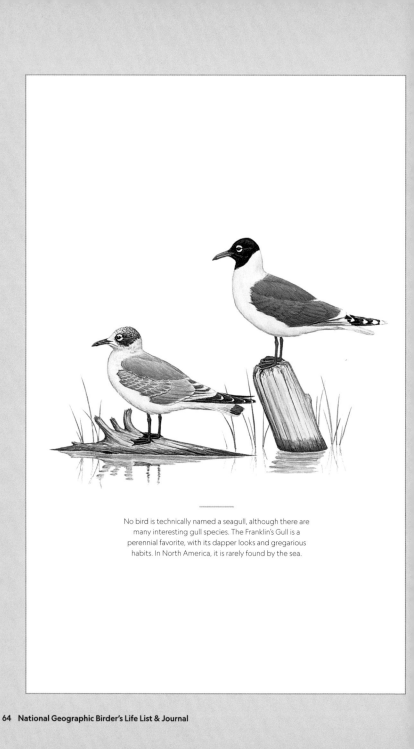

No bird is technically named a seagull, although there are many interesting gull species. The Franklin's Gull is a perennial favorite, with its dapper looks and gregarious habits. In North America, it is rarely found by the sea.

Black-headed Gull *Chroicocephalus ridibundus* **BHGU 2**	When:	Where:
Notes:		

Little Gull *Hydrocoloeus minutus* **LIGU 2**	When:	Where:
Notes:		

Ross's Gull *Rhodostethia rosea* **ROGU 3**	When:	Where:
Notes:		

Laughing Gull *Leucophaeus atricilla* **LAGU 1**	When:	Where:
Notes:		

Franklin's Gull *Leucophaeus pipixcan* **FRGU 1**	When:	Where:
Notes:		

Heermann's Gull *Larus heermanni* HEEG 1	When:	Where:
Notes:		

Common Gull *Larus canus* COGU 3	When:	Where:
Notes:		

Short-billed Gull *Larus brachyrhynchus* SBGU 1	When:	Where:
Notes:		

Ring-billed Gull *Larus delawarensis* RBGU 1	When:	Where:
Notes:		

Western Gull *Larus occidentalis* WEGU 1	When:	Where:
Notes:		

Yellow-footed Gull	When:	Where:
Larus livens YFGU 2		

Notes:

California Gull	When:	Where:
Larus californicus CAGU 1		

Notes:

Herring Gull	When:	Where:
Larus argentatus HERG 1		

Notes:

Iceland Gull	When:	Where:
Larus glaucoides ICGU 1		

Notes:

Lesser Black-backed Gull	When:	Where:
Larus fuscus LBBG 2		

Notes:

Slaty-backed Gull *Larus schistisagus* SBAG 3	When:	Where:
Notes:		

Glaucous-winged Gull *Larus glaucescens* GWGU 1	When:	Where:
Notes:		

Glaucous Gull *Larus hyperboreus* GLGU 1	When:	Where:
Notes:		

Great Black-backed Gull *Larus marinus* GBBG 1	When:	Where:
Notes:		

Brown Noddy *Anous stolidus* BRNO 2	When:	Where:
Notes:		

Black Noddy *Anous minutus* BLNO 2	When:	Where:
Notes:		

Blue-gray Noddy *Anous ceruleus* BGNO 3	When:	Where:
Notes:		

White Tern *Gygis alba* WHTT 2	When:	Where:
Notes:		

Sooty Tern *Onychoprion fuscatus* SOTE 2	When:	Where:
Notes:		

Gray-backed Tern *Onychoprion lunatus* GBAT 2	When:	Where:
Notes:		

Bridled Tern *Onychoprion anaethetus* **BRTE 2**	When:	Where:
Notes:		

Aleutian Tern *Onychoprion aleuticus* **ALTE 2**	When:	Where:
Notes:		

Least Tern *Sternula antillarum* **LETE 1**	When:	Where:
Notes:		

Gull-billed Tern *Gelochelidon nilotica* **GBTE 1**	When:	Where:
Notes:		

Caspian Tern *Hydroprogne caspia* **CATE 1**	When:	Where:
Notes:		

Black Tern *Chlidonias niger* BLTE 1	When:	Where:
Notes:		

Roseate Tern *Sterna dougallii* ROST 2	When:	Where:
Notes:		

Common Tern *Sterna hirundo* COTE 1	When:	Where:
Notes:		

Arctic Tern *Sterna paradisaea* ARTE 1	When:	Where:
Notes:		

Forster's Tern *Sterna forsteri* FOTE 1	When:	Where:
Notes:		

	When:	Where:
Royal Tern *Thalasseus maximus* ROYT 1		
Notes:		

	When:	Where:
Sandwich Tern *Thalasseus sandvicensis* SATE 1		
Notes:		

	When:	Where:
Elegant Tern *Thalasseus elegans* ELTE 1		
Notes:		

	When:	Where:
Black Skimmer *Rynchops niger* BLSK 1		
Notes:		

TROPICBIRDS *(Phaethontidae)*	When:	Where:
White-tailed Tropicbird *Phaethon lepturus* WTTR 2		
Notes:		

Red-billed Tropicbird *Phaethon aethereus* **RBTR 3**	When:	Where:
Notes:		

Red-tailed Tropicbird *Phaethon rubricauda* **RTTR 2**	When:	Where:
Notes:		

LOONS *(Gaviidae)* **Red-throated Loon** *Gavia stellata* **RTLO 1**	When:	Where:
Notes:		

Arctic Loon *Gavia arctica* **ARLO 3**	When:	Where:
Notes:		

Pacific Loon *Gavia pacifica* **PALO 1**	When:	Where:
Notes:		

Common Loon	When:	Where:
Gavia immer COLO 1		
Notes:		

Yellow-billed Loon	When:	Where:
Gavia adamsii YBLO 2		
Notes:		

ALBATROSSES *(Diomedeidae)*	When:	Where:
Laysan Albatross		
Phoebastria immutabilis LAAL 2		
Notes:		

Black-footed Albatross	When:	Where:
Phoebastria nigripes BFAL 1		
Notes:		

Short-tailed Albatross	When:	Where:
Phoebastria albatrus STAL 3		
Notes:		

SOUTHERN STORM-PETRELS (Oceanitidae)	When:	Where:
Wilson's Storm-Petrel *Oceanites oceanicus* **WISP 1**		
Notes:		

	When:	Where:
White-faced Storm-Petrel *Pelagodroma marina* **WFSP 3**		
Notes:		

NORTHERN STORM-PETRELS (Hydrobatidae)	When:	Where:
Fork-tailed Storm-Petrel *Hydrobates furcatus* **FTSP 2**		
Notes:		

	When:	Where:
Leach's Storm-Petrel *Hydrobates leucorhous* **LESP 1**		
Notes:		

	When:	Where:
Townsend's Storm-Petrel *Hydrobates socorroensis* **TOSP 3**		
Notes:		

Ashy Storm-Petrel *Hydrobates homochroa* **ASSP 2**	When:	Where:
Notes:		

Band-rumped Storm-Petrel *Hydrobates castro* **BSTP 2**	When:	Where:
Notes:		

Black Storm-Petrel *Hydrobates melania* **BLSP 2**	When:	Where:
Notes:		

Tristram's Storm-Petrel *Hydrobates tristrami* **TRSP 3**	When:	Where:
Notes:		

Least Storm-Petrel *Hydrobates microsoma* **LSTP 3**	When:	Where:
Notes:		

SHEARWATERS AND PETRELS (Procellariidae)	When:	Where:
Northern Fulmar *Fulmarus glacialis* **NOFU 1**		

Notes:

	When:	Where:
Trindade Petrel *Pterodroma arminjoniana* **TRPE 3**		

Notes:

	When:	Where:
Murphy's Petrel *Pterodroma ultima* **MUPE 3**		

Notes:

	When:	Where:
Mottled Petrel *Pterodroma inexpectata* **MOPE 2**		

Notes:

	When:	Where:
Bermuda Petrel *Pterodroma cahow* **BEPE 3**		

Notes:

Black-capped Petrel *Pterodroma hasitata* **BCPE 2**	When:	Where:
Notes:		

Juan Fernandez Petrel *Pterodroma externa* **JFPE 3**	When:	Where:
Notes:		

Hawaiian Petrel *Pterodroma sandwichensis* **HAPE 2**	When:	Where:
Notes:		

White-necked Petrel *Pterodroma cervicalis* **WNPE 3**	When:	Where:
Notes:		

Bonin Petrel *Pterodroma hypoleuca* **BOPE 3**	When:	Where:
Notes:		

Black-winged Petrel	When:	Where:
Pterodroma nigripennis **BWPE 3**		

Notes:

Fea's Petrel	When:	Where:
Pterodroma feae **FEPE 3**		

Notes:

Cook's Petrel	When:	Where:
Pterodroma cookii **COPE 3**		

Notes:

Bulwer's Petrel	When:	Where:
Bulweria bulwerii **BUPE 3**		

Notes:

Cory's Shearwater	When:	Where:
Calonectris diomedea **CORS 1**		

Notes:

Wedge-tailed Shearwater *Ardenna pacifica* WTSH 2	When:	Where:
Notes:		

Buller's Shearwater *Ardenna bulleri* BULS 2	When:	Where:
Notes:		

Short-tailed Shearwater *Ardenna tenuirostris* STTS 2	When:	Where:
Notes:		

Sooty Shearwater *Ardenna grisea* SOSH 1	When:	Where:
Notes:		

Great Shearwater *Ardenna gravis* GRSH 1	When:	Where:
Notes:		

Virtually the entire population of Great Shearwaters nests
on a couple of remote islands in the South Atlantic and
then disperses thousands of miles for the rest of the year,
sometimes as far as Greenland and the U.S. East Coast.

Pink-footed Shearwater *Ardenna creatopus* PFSH 1	When:	Where:
Notes:		

Flesh-footed Shearwater *Ardenna carneipes* FFSH 3	When:	Where:
Notes:		

Christmas Shearwater *Puffinus nativitatis* CHSH 3	When:	Where:
Notes:		

Manx Shearwater *Puffinus puffinus* MASH 2	When:	Where:
Notes:		

Newell's Shearwater *Puffinus newelli* NESH 2	When:	Where:
Notes:		

Black-vented Shearwater *Puffinus opisthomelas* BVSH 2	When:	Where:
Notes:		

Audubon's Shearwater *Puffinus lherminieri* AUSH 1	When:	Where:
Notes:		

STORKS *(Ciconiidae)* **Wood Stork** *Mycteria americana* WOST 1	When:	Where:
Notes:		

FRIGATEBIRDS *(Fregatidae)* **Magnificent Frigatebird** *Fregata magnificens* MAFR 1	When:	Where:
Notes:		

Great Frigatebird *Fregata minor* GREF 2	When:	Where:
Notes:		

BOOBIES AND GANNETS (Sulidae)	When:	Where:
Masked Booby *Sula dactylatra* **MABO 2**		

Notes:

	When:	Where:
Brown Booby *Sula leucogaster* **BRBO 2**		

Notes:

	When:	Where:
Red-footed Booby *Sula sula* **RFBO 2**		

Notes:

	When:	Where:
Northern Gannet *Morus bassanus* **NOGA 1**		

Notes:

DARTERS (Anhingidae)	When:	Where:
Anhinga *Anhinga anhinga* **ANHI 1**		

Notes:

CORMORANTS (Phalacrocoracidae)	When:	Where:
Brandt's Cormorant *Urile penicillatus* **BRAC 1**		

Notes:

	When:	Where:
Red-faced Cormorant *Urile urile* **RFCO 2**		

Notes:

	When:	Where:
Pelagic Cormorant *Urile pelagicus* **PECO 1**		

Notes:

	When:	Where:
Great Cormorant *Phalacrocorax carbo* **GRCO 1**		

Notes:

	When:	Where:
Double-crested Cormorant *Nannopterum auritum* **DCCO 1**		

Notes:

Neotropic Cormorant *Nannopterum brasilianum* NECO 1	When:	Where:
Notes:		

PELICANS *(Pelecanidae)*	When:	Where:
American White Pelican *Pelecanus erythrorhynchos* AWPE 1		
Notes:		

Brown Pelican *Pelecanus occidentalis* BRPE 1	When:	Where:
Notes:		

BITTERNS, HERONS, AND ALLIES *(Ardeidae)*	When:	Where:
American Bittern *Botaurus lentiginosus* AMBI 1		
Notes:		

Least Bittern *Ixobrychus exilis* LEBI 1	When:	Where:
Notes:		

Great Blue Heron _Ardea herodias_ **GBHE 1**	When:	Where:
Notes:		

Great Egret _Ardea alba_ **GREG 1**	When:	Where:
Notes:		

Snowy Egret _Egretta thula_ **SNEG 1**	When:	Where:
Notes:		

Little Blue Heron _Egretta caerulea_ **LBHE 1**	When:	Where:
Notes:		

Tricolored Heron _Egretta tricolor_ **TRHE 1**	When:	Where:
Notes:		

Reddish Egret *Egretta rufescens* **REEG 1**	When:	Where:
Notes:		

Cattle Egret *Bubulcus ibis* **CAEG 1**	When:	Where:
Notes:		

Green Heron *Butorides virescens* **GRHE 1**	When:	Where:
Notes:		

Black-crowned Night-Heron *Nycticorax nycticorax* **BCNH 1**	When:	Where:
Notes:		

Yellow-crowned Night-Heron *Nyctanassa violacea* **YCNH 1**	When:	Where:
Notes:		

IBISES AND SPOONBILLS (Threskiornithidae)

White Ibis
Eudocimus albus   WHIB 1

When:

Where:

Notes:

Glossy Ibis
Plegadis falcinellus   GLIB 1

When:

Where:

Notes:

White-faced Ibis
Plegadis chihi   WFIB 1

When:

Where:

Notes:

Roseate Spoonbill
Platalea ajaja   ROSP 1

When:

Where:

Notes:

NEW WORLD VULTURES (Cathartidae)

California Condor
Gymnogyps californianus   CACO 2

When:

Where:

Notes:

Black Vulture Coragyps atratus **BLVU 1**	When:	Where:
Notes:		

Turkey Vulture Cathartes aura **TUVU 1**	When:	Where:
Notes:		

OSPREYS (Pandionidae)	When:	Where:
Osprey Pandion haliaetus **OSPR 1**		
Notes:		

HAWKS, KITES, EAGLES, AND ALLIES (Accipitridae)	When:	Where:
White-tailed Kite Elanus leucurus **WTKI 1**		
Notes:		

Hook-billed Kite Chondrohierax uncinatus **HBKI 3**	When:	Where:
Notes:		

Swallow-tailed Kite *Elanoides forficatus* **STKI 1**	When:	Where:
Notes:		

Golden Eagle *Aquila chrysaetos* **GOEA 1**	When:	Where:
Notes:		

Northern Harrier *Circus hudsonius* **NOHA 1**	When:	Where:
Notes:		

Sharp-shinned Hawk *Accipiter striatus* **SSHA 1**	When:	Where:
Notes:		

Cooper's Hawk *Accipiter cooperii* **COHA 1**	When:	Where:
Notes:		

Northern Goshawk *Accipiter gentilis* **NOGO 1**	When:	Where:
Notes:		

Bald Eagle *Haliaeetus leucocephalus* **BAEA 1**	When:	Where:
Notes:		

Mississippi Kite *Ictinia mississippiensis* **MIKI 1**	When:	Where:
Notes:		

Snail Kite *Rostrhamus sociabilis* **SNKI 2**	When:	Where:
Notes:		

Common Black Hawk *Buteogallus anthracinus* **COBH 2**	When:	Where:
Notes:		

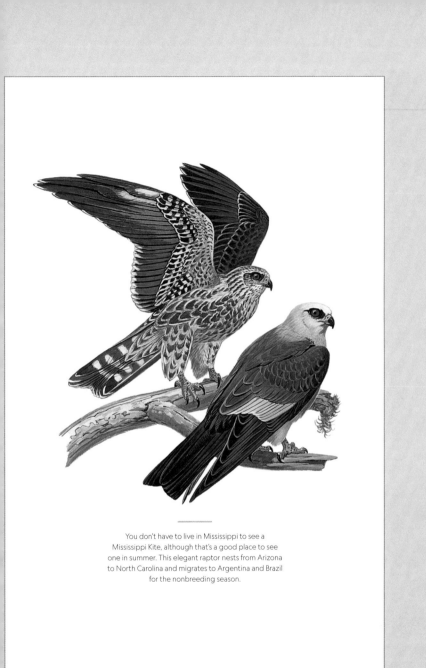

You don't have to live in Mississippi to see a
Mississippi Kite, although that's a good place to see
one in summer. This elegant raptor nests from Arizona
to North Carolina and migrates to Argentina and Brazil
for the nonbreeding season.

Harris's Hawk *Parabuteo unicinctus* **HASH 1**	When:	Where:
Notes:		

White-tailed Hawk *Geranoaetus albicaudatus* **WTHA 2**	When:	Where:
Notes:		

Gray Hawk *Buteo plagiatus* **GRHA 2**	When:	Where:
Notes:		

Red-shouldered Hawk *Buteo lineatus* **RSHA 1**	When:	Where:
Notes:		

Broad-winged Hawk *Buteo platypterus* **BWHA 1**	When:	Where:
Notes:		

| **Hawaiian Hawk** | When: | Where: |
| *Buteo solitarius*   HAWH 2 | | |

Notes:

| **Short-tailed Hawk** | When: | Where: |
| *Buteo brachyurus*   STHA 2 | | |

Notes:

| **Swainson's Hawk** | When: | Where: |
| *Buteo swainsoni*   SWHA 1 | | |

Notes:

| **Zone-tailed Hawk** | When: | Where: |
| *Buteo albonotatus*   ZTHA 2 | | |

Notes:

| **Red-tailed Hawk** | When: | Where: |
| *Buteo jamaicensis*   RTHA 1 | | |

Notes:

Rough-legged Hawk *Buteo lagopus* **RLHA 1**	When:	Where:
Notes:		

Ferruginous Hawk *Buteo regalis* **FEHA 1**	When:	Where:
Notes:		

BARN OWLS *(Tytonidae)* **Barn Owl** *Tyto alba* **BANO 1**	When:	Where:
Notes:		

TYPICAL OWLS *(Strigidae)* **Flammulated Owl** *Psiloscops flammeolus* **FLOW 2**	When:	Where:
Notes:		

Whiskered Screech-Owl *Megascops trichopsis* **WHSO 2**	When:	Where:
Notes:		

Western Screech-Owl *Megascops kennicottii* **WESO 1**	When:	Where:
Notes:		

Eastern Screech-Owl *Megascops asio* **EASO 1**	When:	Where:
Notes:		

Great Horned Owl *Bubo virginianus* **GHOW 1**	When:	Where:
Notes:		

Snowy Owl *Bubo scandiacus* **SNOW 2**	When:	Where:
Notes:		

Northern Hawk Owl *Surnia ulula* **NHOW 2**	When:	Where:
Notes:		

Northern Pygmy-Owl *Glaucidium gnoma* **NOPO 2**	When:	Where:
Notes:		

Ferruginous Pygmy-Owl *Glaucidium brasilianum* **FEPO 3**	When:	Where:
Notes:		

Elf Owl *Micrathene whitneyi* **ELOW 2**	When:	Where:
Notes:		

Burrowing Owl *Athene cunicularia* **BUOW 1**	When:	Where:
Notes:		

Spotted Owl *Strix occidentalis* **SPOW 2**	When:	Where:
Notes:		

Barred Owl *Strix varia* **BADO 1**	When:	Where:
Notes:		

Great Gray Owl *Strix nebulosa* **GGOW 2**	When:	Where:
Notes:		

Long-eared Owl *Asio otus* **LEOW 2**	When:	Where:
Notes:		

Short-eared Owl *Asio flammeus* **SEOW 1**	When:	Where:
Notes:		

Boreal Owl *Aegolius funereus* **BOOW 2**	When:	Where:
Notes:		

	When:	Where:
Northern Saw-whet Owl *Aegolius acadicus* NSWO 2		

Notes:

	When:	Where:
TROGONS *(Trogonidae)* **Elegant Trogon** *Trogon elegans* ELTR 2		

Notes:

	When:	Where:
KINGFISHERS *(Alcedinidae)* **Ringed Kingfisher** *Megaceryle torquata* RIKI 2		

Notes:

	When:	Where:
Belted Kingfisher *Megaceryle alcyon* BEKI 1		

Notes:

	When:	Where:
Green Kingfisher *Chloroceryle americana* GKIN 2		

Notes:

WOODPECKERS AND ALLIES (Picidae)	When:	Where:
Lewis's Woodpecker *Melanerpes lewis* **LEWO 1**		

Notes:

	When:	Where:
Red-headed Woodpecker *Melanerpes erythrocephalus* **RHWO 1**		

Notes:

	When:	Where:
Acorn Woodpecker *Melanerpes formicivorus* **ACWO 1**		

Notes:

	When:	Where:
Gila Woodpecker *Melanerpes uropygialis* **GIWO 1**		

Notes:

	When:	Where:
Golden-fronted Woodpecker *Melanerpes aurifrons* **GFWO 1**		

Notes:

Red-bellied Woodpecker *Melanerpes carolinus* **RBWO 1**	When:	Where:
Notes:		

Williamson's Sapsucker *Sphyrapicus thyroideus* **WISA 1**	When:	Where:
Notes:		

Yellow-bellied Sapsucker *Sphyrapicus varius* **YBSA 1**	When:	Where:
Notes:		

Red-naped Sapsucker *Sphyrapicus nuchalis* **RNSA 1**	When:	Where:
Notes:		

Red-breasted Sapsucker *Sphyrapicus ruber* **RBSA 1**	When:	Where:
Notes:		

American Three-toed Woodpecker *Picoides dorsalis* **ATTW 2**	When:	Where:
Notes:		

Black-backed Woodpecker *Picoides arcticus* **BBWO 2**	When:	Where:
Notes:		

Downy Woodpecker *Dryobates pubescens* **DOWO 1**	When:	Where:
Notes:		

Nuttall's Woodpecker *Dryobates nuttallii* **NUWO 1**	When:	Where:
Notes:		

Ladder-backed Woodpecker *Dryobates scalaris* **LBWO 1**	When:	Where:
Notes:		

Red-cockaded Woodpecker *Dryobates borealis* **RCWO 2**	When:	Where:
Notes:		

Hairy Woodpecker *Dryobates villosus* **HAWO 1**	When:	Where:
Notes:		

White-headed Woodpecker *Dryobates albolarvatus* **WHWO 1**	When:	Where:
Notes:		

Arizona Woodpecker *Dryobates arizonae* **ARWO 2**	When:	Where:
Notes:		

Northern Flicker *Colaptes auratus* **NOFL 1**	When:	Where:
Notes:		

Like other woodpeckers, the Northern Flicker drums on tree trunks and excavates nest cavities, but it also perches on the ground and eats ants. Flickers in eastern North America have golden-yellow wing and tail feathers, while those of western ones are reddish-colored.

| **Gilded Flicker** | When: | Where: |
| *Colaptes chrysoides* **GIFL 2** | | |

Notes:

| **Pileated Woodpecker** | When: | Where: |
| *Dryocopus pileatus* **PIWO 1** | | |

Notes:

CARACARAS AND FALCONS (*Falconidae*)	When:	Where:
Crested Caracara		
Caracara plancus **CRCA 2**		

Notes:

| **American Kestrel** | When: | Where: |
| *Falco sparverius* **AMKE 1** | | |

Notes:

| **Merlin** | When: | Where: |
| *Falco columbarius* **MERL 1** | | |

Notes:

Aplomado Falcon *Falco femoralis* **APFA 3**	When:	Where:
Notes:		

Gyrfalcon *Falco rusticolus* **GYRF 2**	When:	Where:
Notes:		

Peregrine Falcon *Falco peregrinus* **PEFA 1**	When:	Where:
Notes:		

Prairie Falcon *Falco mexicanus* **PRFA 1**	When:	Where:
Notes:		

PARAKEETS, MACAWS, AND PARROTS *(Psittacidae)* **Monk Parakeet** *Myiopsitta monachu* **MOPA 2**	When:	Where:
Notes:		

Nanday Parakeet *Aratinga nenday* NAPA 2	When:	Where:
Notes:		

Green Parakeet *Psittacara holochlorus* GREP 2	When:	Where:
Notes:		

Mitred Parakeet *Psittacara mitratus* MIPA 2	When:	Where:
Notes:		

White-winged Parakeet *Brotogeris versicolurus* WWPA 2	When:	Where:
Notes:		

Yellow-chevroned Parakeet *Brotogeris chiriri* YCPA 2	When:	Where:
Notes:		

Red-crowned Parrot *Amazona viridigenalis* **RCPA 2**	When:	Where:
Notes:		

Rose-ringed Parakeet *Psittacula krameri* **RRPA 2**	When:	Where:
Notes:		

Rosy-faced Lovebird *Agapornis roseicollis* **RFLO 2**	When:	Where:
Notes:		

BECARDS, TITYRAS, AND ALLIES *(Tityridae)*

Rose-throated Becard *Pachyramphus aglaiae* **RTBE 3**	When:	Where:
Notes:		

TYRANT FLYCATCHERS *(Tyrannidae)*

Northern Beardless-Tyrannulet *Camptostoma imberbe* **NOBT 2**	When:	Where:
Notes:		

Dusky-capped Flycatcher *Myiarchus tuberculifer* DCFL 2	When:	Where:
Notes:		

Ash-throated Flycatcher *Myiarchus cinerascens* ATFL 1	When:	Where:
Notes:		

Great Crested Flycatcher *Myiarchus crinitus* GCFL 1	When:	Where:
Notes:		

Brown-crested Flycatcher *Myiarchus tyrannulus* BCFL 1	When:	Where:
Notes:		

La Sagra's Flycatcher *Myiarchus sagrae* LSFL 3	When:	Where:
Notes:		

Great Kiskadee *Pitangus sulphuratus* **GKIS 2**	When:	Where:
Notes:		

Sulphur-bellied Flycatcher *Myiodynastes luteiventris* **SBFL 2**	When:	Where:
Notes:		

Tropical Kingbird *Tyrannus melancholicus* **TRKI 2**	When:	Where:
Notes:		

Couch's Kingbird *Tyrannus couchii* **COKI 2**	When:	Where:
Notes:		

Cassin's Kingbird *Tyrannus vociferans* **CAKI 1**	When:	Where:
Notes:		

Thick-billed Kingbird *Tyrannus crassirostris* **TBKI 2**	When:	Where:
Notes:		

Western Kingbird *Tyrannus verticalis* **WEKI 1**	When:	Where:
Notes:		

Eastern Kingbird *Tyrannus tyrannus* **EAKI 1**	When:	Where:
Notes:		

Gray Kingbird *Tyrannus dominicensis* **GRAK 2**	When:	Where:
Notes:		

Scissor-tailed Flycatcher *Tyrannus forficatus* **STFL 1**	When:	Where:
Notes:		

Fork-tailed Flycatcher *Tyrannus savana* **FTFL 3**	When:	Where:
Notes:		

Olive-sided Flycatcher *Contopus cooperi* **OSFL 1**	When:	Where:
Notes:		

Greater Pewee *Contopus pertinax* **GRPE 2**	When:	Where:
Notes:		

Western Wood-Pewee *Contopus sordidulus* **WEWP 1**	When:	Where:
Notes:		

Eastern Wood-Pewee *Contopus virens* **EAWP 1**	When:	Where:
Notes:		

Yellow-bellied Flycatcher *Empidonax flaviventris* **YBFL 1**	When:	Where:
Notes:		

Acadian Flycatcher *Empidonax virescens* **ACFL 1**	When:	Where:
Notes:		

Alder Flycatcher *Empidonax alnorum* **ALFL 1**	When:	Where:
Notes:		

Willow Flycatcher *Empidonax traillii* **WIFL 1**	When:	Where:
Notes:		

Least Flycatcher *Empidonax minimus* **LEFL 1**	When:	Where:
Notes:		

Hammond's Flycatcher *Empidonax hammondii* **HAFL** 1	When:	Where:
Notes:		

Gray Flycatcher *Empidonax wrightii* **GRFL** 1	When:	Where:
Notes:		

Dusky Flycatcher *Empidonax oberholseri* **DUFL** 1	When:	Where:
Notes:		

Pacific-slope Flycatcher *Empidonax difficilis* **PSFL** 1	When:	Where:
Notes:		

Cordilleran Flycatcher *Empidonax occidentalis* **COFL** 1	When:	Where:
Notes:		

Buff-breasted Flycatcher *Empidonax fulvifrons*　**BBFL 2**	When:	Where:
Notes:		

Black Phoebe *Sayornis nigricans*　**BLPH 1**	When:	Where:
Notes:		

Eastern Phoebe *Sayornis phoebe*　**EAPH 1**	When:	Where:
Notes:		

Say's Phoebe *Sayornis saya*　**SAPH 1**	When:	Where:
Notes:		

Vermilion Flycatcher *Pyrocephalus rubinus*　**VEFL 1**	When:	Where:
Notes:		

VIREOS *(Vireonidae)*	When:	Where:
Black-capped Vireo *Vireo atricapilla* **BCVI 2**		

Notes:

White-eyed Vireo *Vireo griseus* **WEVI 1**	When:	Where:

Notes:

Bell's Vireo *Vireo bellii* **BEVI 1**	When:	Where:

Notes:

Gray Vireo *Vireo vicinior* **GRVI 2**	When:	Where:

Notes:

Hutton's Vireo *Vireo huttoni* **HUVI 1**	When:	Where:

Notes:

Yellow-throated Vireo *Vireo flavifrons* YTVI 1	When:	Where:
Notes:		

Cassin's Vireo *Vireo cassinii* CAVI 1	When:	Where:
Notes:		

Blue-headed Vireo *Vireo solitarius* BHVI 1	When:	Where:
Notes:		

Plumbeous Vireo *Vireo plumbeus* PLVI 1	When:	Where:
Notes:		

Philadelphia Vireo *Vireo philadelphicus* PHVI 1	When:	Where:
Notes:		

Warbling Vireo *Vireo gilvus* WAVI 1	When:	Where:
Notes:		

Red-eyed Vireo *Vireo olivaceus* REVI 1	When:	Where:
Notes:		

Yellow-green Vireo *Vireo flavoviridis* YGVI 3	When:	Where:
Notes:		

Black-whiskered Vireo *Vireo altiloquus* BWVI 2	When:	Where:
Notes:		

MONARCH FLYCATCHERS *(Monarchidae)* **Kauai Elepaio** *Chasiempis sclateri* KAEL 2	When:	Where:
Notes:		

Oahu Elepaio *Chasiempis ibidis* OAEL 2	When:	Where:
Notes:		

Hawaii Elepaio *Chasiempis sandwichensis* HAEL 2	When:	Where:
Notes:		

SHRIKES *(Laniidae)*	When:	Where:
Loggerhead Shrike *Lanius ludovicianus* LOSH 1		
Notes:		

Northern Shrike *Lanius borealis* NSHR 1	When:	Where:
Notes:		

JAYS AND CROWS *(Corvidae)*	When:	Where:
Canada Jay *Perisoreus canadensis* CAJA 1		
Notes:		

Green Jay *Cyanocorax yncas* **GRJA 2**	When:	Where:
Notes:		

Pinyon Jay *Gymnorhinus cyanocephalus* **PIJA 1**	When:	Where:
Notes:		

Steller's Jay *Cyanocitta stelleri* **STJA 1**	When:	Where:
Notes:		

Blue Jay *Cyanocitta cristata* **BLJA 1**	When:	Where:
Notes:		

Florida Scrub-Jay *Aphelocoma coerulescens* **FLSJ 2**	When:	Where:
Notes:		

The Clark's Nutcracker, an icon of craggy pine forests
in the western United States and Canada, readily entertains
mountaineers, backpackers, and high-elevation birders
with its raucous calls and flashy plumage.

Island Scrub-Jay *Aphelocoma insularis* ISSJ 2	When:	Where:
Notes:		

California Scrub-Jay *Aphelocoma californica* CASJ 1	When:	Where:
Notes:		

Woodhouse's Scrub-Jay *Aphelocoma woodhouseii* WOSJ 1	When:	Where:
Notes:		

Mexican Jay *Aphelocoma wollweberi* MEJA 2	When:	Where:
Notes:		

Clark's Nutcracker *Nucifraga columbiana* CLNU 1	When:	Where:
Notes:		

Black-billed Magpie *Pica hudsonia* **BBMA 1**	When:	Where:
Notes:		

Yellow-billed Magpie *Pica nuttalli* **YBMA 2**	When:	Where:
Notes:		

American Crow *Corvus brachyrhynchos* **AMCR 1**	When:	Where:
Notes:		

Fish Crow *Corvus ossifragus* **FICR 1**	When:	Where:
Notes:		

Chihuahuan Raven *Corvus cryptoleucus* **CHRA 1**	When:	Where:
Notes:		

Common Raven _Corvus corax_ **CORA 1**	When:	Where:
Notes:		

VERDINS _(Remizidae)_	When:	Where:
Verdin _Auriparus flaviceps_ **VERD 1**		
Notes:		

CHICKADEES AND TITMICE _(Paridae)_	When:	Where:
Carolina Chickadee _Poecile carolinensis_ **CACH 1**		
Notes:		

Black-capped Chickadee _Poecile atricapillus_ **BCCH 1**	When:	Where:
Notes:		

Mountain Chickadee _Poecile gambeli_ **MOCH 1**	When:	Where:
Notes:		

	When:	Where:
Mexican Chickadee *Poecile sclateri* MECH 2		

Notes:

	When:	Where:
Chestnut-backed Chickadee *Poecile rufescens* CBCH 1		

Notes:

	When:	Where:
Boreal Chickadee *Poecile hudsonicus* BOCH 1		

Notes:

	When:	Where:
Gray-headed Chickadee *Poecile cinctus* GHCH 3		

Notes:

	When:	Where:
Bridled Titmouse *Baeolophus wollweberi* BRTI 2		

Notes:

Oak Titmouse *Baeolophus inornatus* OATI 1	When:	Where:
Notes:		

Juniper Titmouse *Baeolophus ridgwayi* JUTI 1	When:	Where:
Notes:		

Tufted Titmouse *Baeolophus bicolor* TUTI 1	When:	Where:
Notes:		

Black-crested Titmouse *Baeolophus atricristatus* BCTI 2	When:	Where:
Notes:		

LARKS *(Alaudidae)* **Eurasian Skylark** *Alauda arvensis* EUSK 2	When:	Where:
Notes:		

Horned Lark *Eremophila alpestris* **HOLA 1**	When:	Where:
Notes:		

REED WARBLERS (*Acrocephalidae*)	When:	Where:
Millerbird *Acrocephalus familiaris* **MILL 3**		
Notes:		

SWALLOWS (*Hirundinidae*)	When:	Where:
Bank Swallow *Riparia riparia* **BANS 1**		
Notes:		

Tree Swallow *Tachycineta bicolor* **TRES 1**	When:	Where:
Notes:		

Violet-green Swallow *Tachycineta thalassina* **VGSW 1**	When:	Where:
Notes:		

Northern Rough-winged Swallow *Stelgidopteryx serripennis* **NRWS 1**	When:	Where:
Notes:		

Purple Martin *Progne subis* **PUMA 1**	When:	Where:
Notes:		

Barn Swallow *Hirundo rustica* **BARS 1**	When:	Where:
Notes:		

Cliff Swallow *Petrochelidon pyrrhonota* **CLSW 1**	When:	Where:
Notes:		

Cave Swallow *Petrochelidon fulva* **CASW 1**	When:	Where:
Notes:		

BUSHTITS (Aegithalidae)	When:	Where:
Bushtit *Psaltriparus minimus* **BUSH 1**		

Notes:

BUSH-WARBLERS (Cettiidae)	When:	Where:
Japanese Bush-Warbler *Horornis diphone* **JABW 2**		

Notes:

LEAF WARBLERS (Phylloscopidae)	When:	Where:
Arctic Warbler *Phylloscopus borealis* **ARWA 2**		

Notes:

BULBULS (Pycnonotidae)	When:	Where:
Red-vented Bulbul *Pycnonotus cafer* **RVBU 2**		

Notes:

	When:	Where:
Red-whiskered Bulbul *Pycnonotus jocosus* **RWBU 2**		

Notes:

SYLVIID WARBLERS *(Sylviidae)*	When:	Where:
Wrentit *Chamaea fasciata* **WREN 1**		

Notes:

WHITE-EYES *(Zosteropidae)*	When:	Where:
Warbling White-eye *Zosterops japonicus* **WAWE 2**		

Notes:

LAUGHINGTHRUSHES *(Leiothrichidae)*	When:	Where:
Greater Necklaced Laughingthrush *Garrulax pectoralis* **GNLA 2**		

Notes:

	When:	Where:
Hwamei *Garrulax canorus* **HWAM 2**		

Notes:

	When:	Where:
Red-billed Leiothrix *Leiothrix lutea* **RBLE 2**		

Notes:

KINGLETS (Regulidae)	When:	Where:
Ruby-crowned Kinglet *Corthylio calendula* **RCKI 1**		

Notes:

	When:	Where:
Golden-crowned Kinglet *Regulus satrapa* **GCKI 1**		

Notes:

WAXWINGS (Bombycillidae)	When:	Where:
Bohemian Waxwing *Bombycilla garrulus* **BOWA 2**		

Notes:

	When:	Where:
Cedar Waxwing *Bombycilla cedrorum* **CEDW 1**		

Notes:

SILKY-FLYCATCHERS (Ptiliogonatidae)	When:	Where:
Phainopepla *Phainopepla nitens* **PHAI 1**		

Notes:

NUTHATCHES (Sittidae)	When:	Where:
Red-breasted Nuthatch *Sitta canadensis* RBNU 1		

Notes:

	When:	Where:
White-breasted Nuthatch *Sitta carolinensis* WBNU 1		

Notes:

	When:	Where:
Pygmy Nuthatch *Sitta pygmaea* PYNU 1		

Notes:

	When:	Where:
Brown-headed Nuthatch *Sitta pusilla* BHNU 1		

Notes:

CREEPERS (Certhiidae)	When:	Where:
Brown Creeper *Certhia americana* BRCR 1		

Notes:

GNATCATCHERS AND GNATWRENS *(Polioptilidae)*	When:	Where:
Blue-gray Gnatcatcher *Polioptila caerulea* **BGGN 1**		

Notes:

	When:	Where:
Black-tailed Gnatcatcher *Polioptila melanura* **BTGN 1**		

Notes:

	When:	Where:
California Gnatcatcher *Polioptila californica* **CAGN 2**		

Notes:

	When:	Where:
Black-capped Gnatcatcher *Polioptila nigriceps* **BCGN 3**		

Notes:

WRENS *(Troglodytidae)*	When:	Where:
Rock Wren *Salpinctes obsoletus* **ROWR 1**		

Notes:

Canyon Wren *Catherpes mexicanus* **CANW 1**	When:	Where:
Notes:		

House Wren *Troglodytes aedon* **HOWR 1**	When:	Where:
Notes:		

Pacific Wren *Troglodytes pacificus* **PAWR 1**	When:	Where:
Notes:		

Winter Wren *Troglodytes hiemalis* **WIWR 1**	When:	Where:
Notes:		

Sedge Wren *Cistothorus stellaris* **SEWR 1**	When:	Where:
Notes:		

The Marsh Wren boasts a repertoire of dozens, maybe hundreds, of different songs. Eastern (left) and western (right) birds sing differently and may even be separate species. Coastal Marsh Wrens (center), seen from South Carolina to Florida, look more drab.

Marsh Wren *Cistothorus palustris* **MAWR 1**	When:	Where:
Notes:		

Carolina Wren *Thryothorus ludovicianus* **CARW 1**	When:	Where:
Notes:		

Bewick's Wren *Thryomanes bewickii* **BEWR 1**	When:	Where:
Notes:		

Cactus Wren *Campylorhynchus brunneicapillus* **CACW 1**	When:	Where:
Notes:		

MOCKINGBIRDS AND THRASHERS *(Mimidae)* **Gray Catbird** *Dumetella carolinensis* **GRCA 1**	When:	Where:
Notes:		

Curve-billed Thrasher *Toxostoma curvirostre* **CBTH 1**	When:	Where:
Notes:		

Brown Thrasher *Toxostoma rufum* **BRTH 1**	When:	Where:
Notes:		

Long-billed Thrasher *Toxostoma longirostre* **LBTH 2**	When:	Where:
Notes:		

Bendire's Thrasher *Toxostoma bendirei* **BETH 2**	When:	Where:
Notes:		

California Thrasher *Toxostoma redivivum* **CATH 2**	When:	Where:
Notes:		

| **LeConte's Thrasher** | When: | Where: |
| *Toxostoma lecontei* **LCTH 2** | | |

Notes:

| **Crissal Thrasher** | When: | Where: |
| *Toxostoma crissale* **CRTH 2** | | |

Notes:

| **Sage Thrasher** | When: | Where: |
| *Oreoscoptes montanus* **SATH 1** | | |

Notes:

| **Northern Mockingbird** | When: | Where: |
| *Mimus polyglottos* **NOMO 1** | | |

Notes:

STARLINGS *(Sturnidae)*	When:	Where:
European Starling		
Sturnus vulgaris **EUST 1**		

Notes:

Common Myna *Acridotheres tristis* COMY 2	When:	Where:
Notes:		

DIPPERS *(Cinclidae)*	When:	Where:
American Dipper *Cinclus mexicanus* AMDI 1		
Notes:		

THRUSHES *(Turdidae)*	When:	Where:
Eastern Bluebird *Sialia sialis* EABL 1		
Notes:		

Western Bluebird *Sialia mexicana* WEBL 1	When:	Where:
Notes:		

Mountain Bluebird *Sialia currucoides* MOBL 1	When:	Where:
Notes:		

Townsend's Solitaire *Myadestes townsendi* TOSO 1	When:	Where:
Notes:		

Omao *Myadestes obscurus* OMAO 2	When:	Where:
Notes:		

Puaiohi *Myadestes palmeri* PUAI 2	When:	Where:
Notes:		

Veery *Catharus fuscescens* VEER 1	When:	Where:
Notes:		

Gray-cheeked Thrush *Catharus minimus* GCTH 1	When:	Where:
Notes:		

Bicknell's Thrush	When:	Where:
Catharus bicknelli BITH 2		

Notes:

Swainson's Thrush	When:	Where:
Catharus ustulatus SWTH 1		

Notes:

Hermit Thrush	When:	Where:
Catharus guttatus HETH 1		

Notes:

Wood Thrush	When:	Where:
Hylocichla mustelina WOTH 1		

Notes:

Eyebrowed Thrush	When:	Where:
Turdus obscurus EYTH 3		

Notes:

Clay-colored Thrush _Turdus grayi_ **CCTH 2**	When:	Where:
Notes:		

Rufous-backed Robin _Turdus rufopalliatus_ **RBRO 3**	When:	Where:
Notes:		

American Robin _Turdus migratorius_ **AMRO 1**	When:	Where:
Notes:		

Varied Thrush _Ixoreus naevius_ **VATH 1**	When:	Where:
Notes:		

OLD WORLD FLYCATCHERS _(Muscicapidae)_ **White-rumped Shama** _Copsychus malabaricus_ **WRSH 2**	When:	Where:
Notes:		

Bluethroat *Cyanecula svecica* BLUE 2	When:	Where:
Notes:		

Siberian Rubythroat *Calliope calliope* SIRU 3	When:	Where:
Notes:		

Northern Wheatear *Oenanthe oenanthe* NOWH 2	When:	Where:
Notes:		

OLIVE WARBLERS *(Peucedramidae)*	When:	Where:
Olive Warbler *Peucedramus taeniatus* OLWA 2		
Notes:		

WAXBILLS *(Estrildidae)*	When:	Where:
African Silverbill *Euodice cantans* AFSI 2		
Notes:		

Java Sparrow *Padda oryzivora* **JASP** 2	When:	Where:
Notes:		

Scaly-breasted Munia *Lonchura punctulata* **SBMU** 2	When:	Where:
Notes:		

Chestnut Munia *Lonchura atricapilla* **CHMU** 2	When:	Where:
Notes:		

Red Avadavat *Amandava amandava* **REAV** 2	When:	Where:
Notes:		

Common Waxbill *Estrilda astrild* **COMW** 2	When:	Where:
Notes:		

OLD WORLD SPARROWS (Passeridae)	When:	Where:
House Sparrow *Passer domesticus* **HOSP 1**		

Notes:

	When:	Where:
Eurasian Tree Sparrow *Passer montanus* **ETSP 2**		

Notes:

WAGTAILS AND PIPITS (Motacillidae)	When:	Where:
Eastern Yellow Wagtail *Motacilla tschutschensis* **EYWA 2**		

Notes:

	When:	Where:
White Wagtail *Motacilla alba* **WHWA 3**		

Notes:

	When:	Where:
Olive-backed Pipit *Anthus hodgsoni* **OBPI 3**		

Notes:

Red-throated Pipit *Anthus cervinus* RTPI 3	When:	Where:
Notes:		

American Pipit *Anthus rubescens* AMPI 1	When:	Where:
Notes:		

Sprague's Pipit *Anthus spragueii* SPPI 2	When:	Where:
Notes:		

FRINGILLINE AND CARDUELINE FINCHES AND ALLIES *(Fringillidae)* **Brambling** *Fringilla montifringilla* BRAM 3	When:	Where:
Notes:		

Evening Grosbeak *Coccothraustes vespertinus* EVGR 1	When:	Where:
Notes:		

Akikiki	When:	Where:
Oreomystis bairdi **AKIK 2**		

Notes:

Maui Alauahio	When:	Where:
Paroreomyza montana **MAAL 2**		

Notes:

Palila	When:	Where:
Loxioides bailleui **PALI 2**		

Notes:

Laysan Finch	When:	Where:
Telespiza cantans **LAFI 3**		

Notes:

Nihoa Finch	When:	Where:
Telespiza ultima **NIFI 3**		

Notes:

Akohekohe	When:	Where:
Palmeria dolei **AKOH 2**		

Notes:

Apapane	When:	Where:
Himatione sanguinea **APAP 2**		

Notes:

Iiwi	When:	Where:
Drepanis coccinea **IIWI 2**		

Notes:

Maui Parrotbill	When:	Where:
Pseudonestor xanthophrys **MAPA 2**		

Notes:

Akiapolaau	When:	Where:
Hemignathus wilsoni **AKIA 2**		

Notes:

Anianiau	When:	Where:
Magumma parva **ANIA 2**		

Notes:

Hawaii Amakihi	When:	Where:
Chlorodrepanis virens **HAAM 2**		

Notes:

Oahu Amakihi	When:	Where:
Chlorodrepanis flava **OAAM 2**		

Notes:

Kauai Amakihi	When:	Where:
Chlorodrepanis stejnegeri **KAAM 2**		

Notes:

Hawaii Creeper	When:	Where:
Loxops mana **HCRE 2**		

Notes:

Akekee	When:	Where:
Loxops caeruleirostris **AKEK 2**		

Notes:

Hawaii Akepa	When:	Where:
Loxops coccineus **HAAK 2**		

Notes:

Pine Grosbeak	When:	Where:
Pinicola enucleator **PIGR 1**		

Notes:

Gray-crowned Rosy-Finch	When:	Where:
Leucosticte tephrocotis **GCRF 1**		

Notes:

Black Rosy-Finch	When:	Where:
Leucosticte atrata **BLRF 2**		

Notes:

Brown-capped Rosy-Finch *Leucosticte australis* **BCRF 2**	When:	Where:
Notes:		

House Finch *Haemorhous mexicanus* **HOFI 1**	When:	Where:
Notes:		

Purple Finch *Haemorhous purpureus* **PUFI 1**	When:	Where:
Notes:		

Cassin's Finch *Haemorhous cassinii* **CAFI 1**	When:	Where:
Notes:		

Yellow-fronted Canary *Crithagra mozambica* **YFCA 2**	When:	Where:
Notes:		

Common Redpoll *Acanthis flammea* **CORE 1**	When:	Where:
Notes:		

Hoary Redpoll *Acanthis hornemanni* **HORE 2**	When:	Where:
Notes:		

Red Crossbill *Loxia curvirostra* **RECR 1**	When:	Where:
Notes:		

Cassia Crossbill *Loxia sinesciuris* **CACR 2**	When:	Where:
Notes:		

White-winged Crossbill *Loxia leucoptera* **WWCR 2**	When:	Where:
Notes:		

How did the American Goldfinch become the official
state bird of Washington, Iowa, and New Jersey?
Its sunny appearance and cheerful personality
surely helped. This little bird brightens our days
from coast to coast.

Pine Siskin *Spinus pinus* PISI 1	When:	Where:
Notes:		

Lesser Goldfinch *Spinus psaltria* LEGO 1	When:	Where:
Notes:		

Lawrence's Goldfinch *Spinus lawrencei* LAGO 2	When:	Where:
Notes:		

American Goldfinch *Spinus tristis* AMGO 1	When:	Where:
Notes:		

Island Canary *Serinus canaria* ISCA 3	When:	Where:
Notes:		

LONGSPURS AND SNOW BUNTINGS (Calcariidae)	When:	Where:
Lapland Longspur *Calcarius lapponicus* **LALO 1**		

Notes:

	When:	Where:
Chestnut-collared Longspur *Calcarius ornatus* **CCLO 1**		

Notes:

	When:	Where:
Smith's Longspur *Calcarius pictus* **SMLO 2**		

Notes:

	When:	Where:
Thick-billed Longspur *Rhynchophanes mccownii* **TBLO 2**		

Notes:

	When:	Where:
Snow Bunting *Plectrophenax nivalis* **SNBU 1**		

Notes:

| **McKay's Bunting** | When: | Where: |
| *Plectrophenax hyperboreus*   **MKBU 3** | | |

Notes:

EMBERIZIDS *(Emberizidae)*

| **Rustic Bunting** | When: | Where: |
| *Emberiza rustica*   **RUBU 3** | | |

Notes:

TOWHEES AND SPARROWS *(Passerellidae)*

| **Rufous-winged Sparrow** | When: | Where: |
| *Peucaea carpalis*   **RWSP 2** | | |

Notes:

| **Botteri's Sparrow** | When: | Where: |
| *Peucaea botterii*   **BOSP 2** | | |

Notes:

| **Cassin's Sparrow** | When: | Where: |
| *Peucaea cassinii*   **CASP 1** | | |

Notes:

Bachman's Sparrow *Peucaea aestivalis* **BACS 1**	When:	Where:
Notes:		

Grasshopper Sparrow *Ammodramus savannarum* **GRSP 1**	When:	Where:
Notes:		

Olive Sparrow *Arremonops rufivirgatus* **OLSP 2**	When:	Where:
Notes:		

Five-striped Sparrow *Amphispizopsis quinquestriata* **FSSP 3**	When:	Where:
Notes:		

Black-throated Sparrow *Amphispiza bilineata* **BTSP 1**	When:	Where:
Notes:		

Lark Sparrow *Chondestes grammacus* **LASP 1**	When:	Where:
Notes:		

Lark Bunting *Calamospiza melanocorys* **LARB 1**	When:	Where:
Notes:		

Chipping Sparrow *Spizella passerina* **CHSP 1**	When:	Where:
Notes:		

Clay-colored Sparrow *Spizella pallida* **CCSP 1**	When:	Where:
Notes:		

Black-chinned Sparrow *Spizella atrogularis* **BCSP 1**	When:	Where:
Notes:		

Field Sparrow *Spizella pusilla* **FISP 1**	When:	Where:
Notes:		

Brewer's Sparrow *Spizella breweri* **BRSP 1**	When:	Where:
Notes:		

Fox Sparrow *Passerella iliaca* **FOSP 1**	When:	Where:
Notes:		

American Tree Sparrow *Spizelloides arborea* **ATSP 1**	When:	Where:
Notes:		

Dark-eyed Junco *Junco hyemalis* **DEJU 1**	When:	Where:
Notes:		

Yellow-eyed Junco *Junco phaeonotus* **YEJU 2**	When:	Where:
Notes:		

White-crowned Sparrow *Zonotrichia leucophrys* **WCSP 1**	When:	Where:
Notes:		

Golden-crowned Sparrow *Zonotrichia atricapilla* **GCSP 1**	When:	Where:
Notes:		

Harris's Sparrow *Zonotrichia querula* **HASP 1**	When:	Where:
Notes:		

White-throated Sparrow *Zonotrichia albicollis* **WTSP 1**	When:	Where:
Notes:		

Sagebrush Sparrow *Artemisiospiza nevadensis* **SABS 1**	When:	Where:
Notes:		

Bell's Sparrow *Artemisiospiza belli* **BESP 1**	When:	Where:
Notes:		

Vesper Sparrow *Pooecetes gramineus* **VESP 1**	When:	Where:
Notes:		

LeConte's Sparrow *Ammospiza leconteii* **LCSP 1**	When:	Where:
Notes:		

Seaside Sparrow *Ammospiza maritima* **SESP 1**	When:	Where:
Notes:		

Nelson's Sparrow *Ammospiza nelsoni* **NESP 1**	When:	Where:
Notes:		

Saltmarsh Sparrow *Ammospiza caudacuta* **SALS 1**	When:	Where:
Notes:		

Baird's Sparrow *Centronyx bairdii* **BAIS 2**	When:	Where:
Notes:		

Henslow's Sparrow *Centronyx henslowii* **HESP 2**	When:	Where:
Notes:		

Savannah Sparrow *Passerculus sandwichensis* **SAVS 1**	When:	Where:
Notes:		

Song Sparrow *Melospiza melodia* SOSP 1	When:	Where:
Notes:		

Lincoln's Sparrow *Melospiza lincolnii* LISP 1	When:	Where:
Notes:		

Swamp Sparrow *Melospiza georgiana* SWSP 1	When:	Where:
Notes:		

Canyon Towhee *Melozone fusca* CANT 1	When:	Where:
Notes:		

Abert's Towhee *Melozone aberti* ABTO 1	When:	Where:
Notes:		

Marshes, bogs, wet meadows, and other quagmires
may harbor the skulking Swamp Sparrow, more often heard
than seen. Listen for its sharp *chink* calls and, in the summer,
a clear and steady trilling song.

California Towhee *Melozone crissalis* **CALT 1**	When:	Where:
Notes:		

Rufous-crowned Sparrow *Aimophila ruficeps* **RCSP 1**	When:	Where:
Notes:		

Green-tailed Towhee *Pipilo chlorurus* **GTTO 1**	When:	Where:
Notes:		

Spotted Towhee *Pipilo maculatus* **SPTO 1**	When:	Where:
Notes:		

Eastern Towhee *Pipilo erythrophthalmus* **EATO 1**	When:	Where:
Notes:		

SPINDALISES (Spindalidae)

When:

Where:

Western Spindalis
Spindalis zena **WESP 3**

Notes:

YELLOW-BREASTED CHATS (Icteriidae)

When:

Where:

Yellow-breasted Chat
Icteria virens **YBCH 1**

Notes:

BLACKBIRDS (Icteridae)

When:

Where:

Yellow-headed Blackbird
Xanthocephalus xanthocephalus **YHBL 1**

Notes:

When:

Where:

Bobolink
Dolichonyx oryzivorus **BOBO 1**

Notes:

When:

Where:

Eastern Meadowlark
Sturnella magna **EAME 1**

Notes:

Western Meadowlark *Sturnella neglecta* **WEME 1**	When:	Where:
Notes:		

Orchard Oriole *Icterus spurius* **OROR 1**	When:	Where:
Notes:		

Hooded Oriole *Icterus cucullatus* **HOOR 1**	When:	Where:
Notes:		

Bullock's Oriole *Icterus bullockii* **BUOR 1**	When:	Where:
Notes:		

Spot-breasted Oriole *Icterus pectoralis* **SBOR 2**	When:	Where:
Notes:		

Altamira Oriole *Icterus gularis* ALOR 2	When:	Where:
Notes:		

Audubon's Oriole *Icterus graduacauda* AUOR 2	When:	Where:
Notes:		

Baltimore Oriole *Icterus galbula* BAOR 1	When:	Where:
Notes:		

Scott's Oriole *Icterus parisorum* SCOR 1	When:	Where:
Notes:		

Red-winged Blackbird *Agelaius phoeniceus* RWBL 1	When:	Where:
Notes:		

Tricolored Blackbird *Agelaius tricolor* TRBL 2	When:	Where:
Notes:		

Shiny Cowbird *Molothrus bonariensis* SHCO 3	When:	Where:
Notes:		

Bronzed Cowbird *Molothrus aeneus* BROC 1	When:	Where:
Notes:		

Brown-headed Cowbird *Molothrus ater* BHCO 1	When:	Where:
Notes:		

Rusty Blackbird *Euphagus carolinus* RUBL 1	When:	Where:
Notes:		

Brewer's Blackbird *Euphagus cyanocephalus* BRBL 1	When:	Where:

Notes:

Common Grackle *Quiscalus quiscula* COGR 1	When:	Where:

Notes:

Boat-tailed Grackle *Quiscalus major* BTGR 1	When:	Where:

Notes:

Great-tailed Grackle *Quiscalus mexicanus* GTGR 1	When:	Where:

Notes:

WOOD-WARBLERS *(Parulidae)*

Ovenbird *Seiurus aurocapilla* OVEN 1	When:	Where:

Notes:

Worm-eating Warbler *Helmitheros vermivorum* **WEWA 1**	When:	Where:
Notes:		

Louisiana Waterthrush *Parkesia motacilla* **LOWA 1**	When:	Where:
Notes:		

Northern Waterthrush *Parkesia noveboracensis* **NOWA 1**	When:	Where:
Notes:		

Golden-winged Warbler *Vermivora chrysoptera* **GWWA 2**	When:	Where:
Notes:		

Blue-winged Warbler *Vermivora cyanoptera* **BWWA 1**	When:	Where:
Notes:		

Black-and-white Warbler *Mniotilta varia* **BAWW 1**	When:	Where:

Notes:

Prothonotary Warbler *Protonotaria citrea* **PROW 1**	When:	Where:

Notes:

Swainson's Warbler *Limnothlypis swainsonii* **SWWA 2**	When:	Where:

Notes:

Tennessee Warbler *Leiothlypis peregrina* **TEWA 1**	When:	Where:

Notes:

Orange-crowned Warbler *Leiothlypis celata* **OCWA 1**	When:	Where:

Notes:

	When:	Where:
Colima Warbler *Leiothlypis crissalis* **COLW 2**		

Notes:

	When:	Where:
Lucy's Warbler *Leiothlypis luciae* **LUWA 1**		

Notes:

	When:	Where:
Nashville Warbler *Leiothlypis ruficapilla* **NAWA 1**		

Notes:

	When:	Where:
Virginia's Warbler *Leiothlypis virginiae* **VIWA 1**		

Notes:

	When:	Where:
Connecticut Warbler *Oporornis agilis* **CONW 2**		

Notes:

MacGillivray's Warbler *Geothlypis tolmiei* **MGWA 1**	When:	Where:
Notes:		

Mourning Warbler *Geothlypis philadelphia* **MOWA 1**	When:	Where:
Notes:		

Kentucky Warbler *Geothlypis formosa* **KEWA 1**	When:	Where:
Notes:		

Common Yellowthroat *Geothlypis trichas* **COYE 1**	When:	Where:
Notes:		

Hooded Warbler *Setophaga citrina* **HOWA 1**	When:	Where:
Notes:		

American Redstart *Setophaga ruticilla* **AMRE 1**	When:	Where:
Notes:		

Kirtland's Warbler *Setophaga kirtlandii* **KIWA 2**	When:	Where:
Notes:		

Cape May Warbler *Setophaga tigrina* **CMWA 1**	When:	Where:
Notes:		

Cerulean Warbler *Setophaga cerulea* **CERW 2**	When:	Where:
Notes:		

Northern Parula *Setophaga americana* **NOPA 1**	When:	Where:
Notes:		

Tropical Parula *Setophaga pitiayumi* **TRPA 3**	When:	Where:
Notes:		

Magnolia Warbler *Setophaga magnolia* **MAWA 1**	When:	Where:
Notes:		

Bay-breasted Warbler *Setophaga castanea* **BBWA 1**	When:	Where:
Notes:		

Blackburnian Warbler *Setophaga fusca* **BLBW 1**	When:	Where:
Notes:		

Yellow Warbler *Setophaga petechia* **YEWA 1**	When:	Where:
Notes:		

Chestnut-sided Warbler *Setophaga pensylvanica* **CSWA 1**	When:	Where:
Notes:		

Blackpoll Warbler *Setophaga striata* **BLPW 1**	When:	Where:
Notes:		

Black-throated Blue Warbler *Setophaga caerulescens* **BTBW 1**	When:	Where:
Notes:		

Palm Warbler *Setophaga palmarum* **PAWA 1**	When:	Where:
Notes:		

Pine Warbler *Setophaga pinus* **PIWA 1**	When:	Where:
Notes:		

Yellow-rumped Warbler *Setophaga coronata* **YRWA 1**	When:	Where:
Notes:		

Yellow-throated Warbler *Setophaga dominica* **YTWA 1**	When:	Where:
Notes:		

Prairie Warbler *Setophaga discolor* **PRAW 1**	When:	Where:
Notes:		

Grace's Warbler *Setophaga graciae* **GRWA 1**	When:	Where:
Notes:		

Black-throated Gray Warbler *Setophaga nigrescens* **BTYW 1**	When:	Where:
Notes:		

Townsend's Warbler *Setophaga townsendi* **TOWA 1**	When:	Where:
Notes:		

Hermit Warbler *Setophaga occidentalis* **HEWA 1**	When:	Where:
Notes:		

Golden-cheeked Warbler *Setophaga chrysoparia* **GCWA 2**	When:	Where:
Notes:		

Black-throated Green Warbler *Setophaga virens* **BTNW 1**	When:	Where:
Notes:		

Rufous-capped Warbler *Basileuterus rufifrons* **RCWA 3**	When:	Where:
Notes:		

| **Canada Warbler** | When: | Where: |
| *Cardellina canadensis* **CAWA 1** | | |

Notes:

| **Wilson's Warbler** | When: | Where: |
| *Cardellina pusilla* **WIWA 1** | | |

Notes:

| **Red-faced Warbler** | When: | Where: |
| *Cardellina rubrifrons* **RFWA 2** | | |

Notes:

| **Painted Redstart** | When: | Where: |
| *Myioborus pictus* **PARE 2** | | |

Notes:

CARDINALS, PIRANGA TANAGERS, AND ALLIES *(Cardinalidae)*	When:	Where:
Hepatic Tanager		
Piranga flava **HETA 2**		

Notes:

Summer Tanager *Piranga rubra* SUTA 1	When:	Where:
Notes:		

Scarlet Tanager *Piranga olivacea* SCTA 1	When:	Where:
Notes:		

Western Tanager *Piranga ludoviciana* WETA 1	When:	Where:
Notes:		

Flame-colored Tanager *Piranga bidentata* FCTA 3	When:	Where:
Notes:		

Northern Cardinal *Cardinalis cardinalis* NOCA 1	When:	Where:
Notes:		

Pyrrhuloxia *Cardinalis sinuatus* **PYRR 1**	When:	Where:
Notes:		

Rose-breasted Grosbeak *Pheucticus ludovicianus* **RBGR 1**	When:	Where:
Notes:		

Black-headed Grosbeak *Pheucticus melanocephalus* **BHGR 1**	When:	Where:
Notes:		

Blue Grosbeak *Passerina caerulea* **BLGR 1**	When:	Where:
Notes:		

Lazuli Bunting *Passerina amoena* **LAZB 1**	When:	Where:
Notes:		

Indigo Bunting *Passerina cyanea* **INBU 1**	When:	Where:
Notes:		

Varied Bunting *Passerina versicolor* **VABU 2**	When:	Where:
Notes:		

Painted Bunting *Passerina ciris* **PABU 1**	When:	Where:
Notes:		

Dickcissel *Spiza americana* **DICK 1**	When:	Where:
Notes:		

TANAGERS AND ALLIES *(Thraupidae)* **Red-crested Cardinal** *Paroaria coronata* **RCCA 2**	When:	Where:
Notes:		

Yellow-billed Cardinal *Paroaria capitata* **YBCA 2**	When:	Where:
Notes:		

Saffron Finch *Sicalis flaveola* **SAFI 2**	When:	Where:
Notes:		

Yellow-faced Grassquit *Tiaris olivaceus* **YFGR 3**	When:	Where:
Notes:		

Morelet's Seedeater *Sporophila morelleti* **MOSE 3**	When:	Where:
Notes:		

The diminutive Red-breasted Nuthatch inhabits
northern conifer forests. In winter, however,
these hardy birds may travel far south of their usual
range—apparently driven by cone crops—and may
appear all across Canada and the United States.

| CHECKLIST

This checklist can be used to remember the dates on which you saw the same species—or birds found on trips, ones seen in different years, or any other lists you'd like to keep. Go wild! Why not start a column for "birds heard vocalizing" or "birds observed at night"? A blank checklist is full of possibilities. ∎

DUCKS, GEESE, AND SWANS						
Black-bellied Whistling-Duck						
Fulvous Whistling-Duck						
Emperor Goose						
Snow Goose						
Ross's Goose						
Greater White-fronted Goose						
Taiga Bean-Goose						
Tundra Bean-Goose						
Brant						
Cackling Goose						
Canada Goose						
Hawaiian Goose						
Mute Swan						
Trumpeter Swan						
Tundra Swan						
Whooper Swan						
Egyptian Goose						
Muscovy Duck						

Wood Duck						
Blue-winged Teal						
Cinnamon Teal						
Northern Shoveler						
Gadwall						
Eurasian Wigeon						
American Wigeon						
Laysan Duck						
Hawaiian Duck						
Mallard						
Mexican Duck						
American Black Duck						
Mottled Duck						
Northern Pintail						
Green-winged Teal						
Canvasback						
Redhead						
Common Pochard						
Ring-necked Duck						
Tufted Duck						
Greater Scaup						
Lesser Scaup						
Steller's Eider						
Spectacled Eider						
King Eider						

Common Eider						
Harlequin Duck						
Surf Scoter						
White-winged Scoter						
Stejneger's Scoter						
Black Scoter						
Long-tailed Duck						
Bufflehead						
Common Goldeneye						
Barrow's Goldeneye						
Smew						
Hooded Merganser						
Common Merganser						
Red-breasted Merganser						
Masked Duck						
Ruddy Duck						
CURASSOWS AND GUANS						
Plain Chachalaca						
NEW WORLD QUAIL						
Mountain Quail						
Northern Bobwhite						
Scaled Quail						
California Quail						
Gambel's Quail						
Montezuma Quail						

PARTRIDGES, GROUSE, TURKEYS, AND OLD WORLD QUAIL							
Wild Turkey							
Ruffed Grouse							
Spruce Grouse							
Willow Ptarmigan							
Rock Ptarmigan							
White-tailed Ptarmigan							
Greater Sage-Grouse							
Gunnison Sage-Grouse							
Dusky Grouse							
Sooty Grouse							
Sharp-tailed Grouse							
Greater Prairie-Chicken							
Lesser Prairie-Chicken							
Gray Partridge							
Ring-necked Pheasant							
Kalij Pheasant							
Indian Peafowl							
Gray Francolin							
Black Francolin							
Red Junglefowl							
Himalayan Snowcock							
Chukar							
Erckel's Francolin							

FLAMINGOS						
American Flamingo						
GREBES						
Least Grebe						
Pied-billed Grebe						
Horned Grebe						
Red-necked Grebe						
Eared Grebe						
Western Grebe						
Clark's Grebe						
SANDGROUSES						
Chestnut-bellied Sandgrouse						
PIGEONS AND DOVES						
Rock Pigeon						
White-crowned Pigeon						
Red-billed Pigeon						
Band-tailed Pigeon						
Eurasian Collared-Dove						
Spotted Dove						
Zebra Dove						
Inca Dove						
Common Ground Dove						
Ruddy Ground Dove						
White-tipped Dove						
White-winged Dove						

Mourning Dove						
CUCKOOS, ROADRUNNERS, AND ANIS						
Smooth-billed Ani						
Groove-billed Ani						
Greater Roadrunner						
Common Cuckoo						
Yellow-billed Cuckoo						
Mangrove Cuckoo						
Black-billed Cuckoo						
GOATSUCKERS						
Lesser Nighthawk						
Common Nighthawk						
Antillean Nighthawk						
Common Pauraque						
Common Poorwill						
Chuck-will's-widow						
Buff-collared Nightjar						
Eastern Whip-poor-will						
Mexican Whip-poor-will						
SWIFTS						
Black Swift						
Chimney Swift						
Vaux's Swift						
Mariana Swiftlet						
White-throated Swift						

HUMMINGBIRDS						
Mexican Violetear						
Rivoli's Hummingbird						
Blue-throated Mountain-gem						
Lucifer Hummingbird						
Ruby-throated Hummingbird						
Black-chinned Hummingbird						
Anna's Hummingbird						
Costa's Hummingbird						
Calliope Hummingbird						
Rufous Hummingbird						
Allen's Hummingbird						
Broad-tailed Hummingbird						
Broad-billed Hummingbird						
White-eared Hummingbird						
Violet-crowned Hummingbird						
Berylline Hummingbird						
Buff-bellied Hummingbird						
RAILS, GALLINULES, AND COOTS						
Ridgway's Rail						
Clapper Rail						
King Rail						
Virginia Rail						
Sora						
Common Gallinule						

Hawaiian Coot					
American Coot					
Purple Gallinule					
Purple Swamphen					
Yellow Rail					
Black Rail					
LIMPKINS					
Limpkin					
CRANES					
Sandhill Crane					
Whooping Crane					
STILTS AND AVOCETS					
Black-necked Stilt					
American Avocet					
OYSTERCATCHERS					
American Oystercatcher					
Black Oystercatcher					
LAPWINGS AND PLOVERS					
Black-bellied Plover					
American Golden-Plover					
Pacific Golden-Plover					
Killdeer					
Common Ringed Plover					
Semipalmated Plover					

Piping Plover						
Lesser Sand-Plover						
Wilson's Plover						
Mountain Plover						
Snowy Plover						
SANDPIPERS, PHALAROPES, AND ALLIES						
Upland Sandpiper						
Bristle-thighed Curlew						
Whimbrel						
Long-billed Curlew						
Bar-tailed Godwit						
Black-tailed Godwit						
Hudsonian Godwit						
Marbled Godwit						
Ruddy Turnstone						
Black Turnstone						
Red Knot						
Surfbird						
Ruff						
Sharp-tailed Sandpiper						
Stilt Sandpiper						
Curlew Sandpiper						
Temminck's Stint						
Long-toed Stint						
Red-necked Stint						

Sanderling						
Dunlin						
Rock Sandpiper						
Purple Sandpiper						
Baird's Sandpiper						
Least Sandpiper						
White-rumped Sandpiper						
Buff-breasted Sandpiper						
Pectoral Sandpiper						
Semipalmated Sandpiper						
Western Sandpiper						
Short-billed Dowitcher						
Long-billed Dowitcher						
American Woodcock						
Common Snipe						
Wilson's Snipe						
Terek Sandpiper						
Common Sandpiper						
Spotted Sandpiper						
Solitary Sandpiper						
Gray-tailed Tattler						
Wandering Tattler						
Lesser Yellowlegs						
Willet						
Common Greenshank						

Greater Yellowlegs						
Wood Sandpiper						
Wilson's Phalarope						
Red-necked Phalarope						
Red Phalarope						
SKUAS AND JAEGERS						
Great Skua						
South Polar Skua						
Pomarine Jaeger						
Parasitic Jaeger						
Long-tailed Jaeger						
AUKS, MURRES, AND PUFFINS						
Dovekie						
Common Murre						
Thick-billed Murre						
Razorbill						
Black Guillemot						
Pigeon Guillemot						
Long-billed Murrelet						
Marbled Murrelet						
Kittlitz's Murrelet						
Scripps's Murrelet						
Guadalupe Murrelet						
Craveri's Murrelet						
Ancient Murrelet						

Cassin's Auklet						
Parakeet Auklet						
Least Auklet						
Whiskered Auklet						
Crested Auklet						
Rhinoceros Auklet						
Atlantic Puffin						
Horned Puffin						
Tufted Puffin						
GULLS, TERNS, AND SKIMMERS						
Black-legged Kittiwake						
Red-legged Kittiwake						
Ivory Gull						
Sabine's Gull						
Bonaparte's Gull						
Black-headed Gull						
Little Gull						
Ross's Gull						
Laughing Gull						
Franklin's Gull						
Heermann's Gull						
Common Gull						
Short-billed Gull						
Ring-billed Gull						
Western Gull						

Yellow-footed Gull						
California Gull						
Herring Gull						
Iceland Gull						
Lesser Black-backed Gull						
Slaty-backed Gull						
Glaucous-winged Gull						
Glaucous Gull						
Great Black-backed Gull						
Brown Noddy						
Black Noddy						
Blue-gray Noddy						
White Tern						
Sooty Tern						
Gray-backed Tern						
Bridled Tern						
Aleutian Tern						
Least Tern						
Gull-billed Tern						
Caspian Tern						
Black Tern						
Roseate Tern						
Common Tern						
Arctic Tern						
Forster's Tern						

Royal Tern						
Sandwich Tern						
Elegant Tern						
Black Skimmer						
TROPICBIRDS						
White-tailed Tropicbird						
Red-billed Tropicbird						
Red-tailed Tropicbird						
LOONS						
Red-throated Loon						
Arctic Loon						
Pacific Loon						
Common Loon						
Yellow-billed Loon						
ALBATROSSES						
Laysan Albatross						
Black-footed Albatross						
Short-tailed Albatross						
SOUTHERN STORM-PETRELS						
Wilson's Storm-Petrel						
White-faced Storm-Petrel						
NORTHERN STORM-PETRELS						
Fork-tailed Storm-Petrel						
Leach's Storm-Petrel						
Townsend's Storm-Petrel						

Ashy Storm-Petrel						
Band-rumped Storm-Petrel						
Black Storm-Petrel						
Tristram's Storm-Petrel						
Least Storm-Petrel						
SHEARWATERS AND PETRELS						
Northern Fulmar						
Trindade Petrel						
Murphy's Petrel						
Mottled Petrel						
Bermuda Petrel						
Black-capped Petrel						
Juan Fernandez Petrel						
Hawaiian Petrel						
White-necked Petrel						
Bonin Petrel						
Black-winged Petrel						
Fea's Petrel						
Cook's Petrel						
Bulwer's Petrel						
Cory's Shearwater						
Wedge-tailed Shearwater						
Buller's Shearwater						
Short-tailed Shearwater						
Sooty Shearwater						

Great Shearwater						
Pink-footed Shearwater						
Flesh-footed Shearwater						
Christmas Shearwater						
Manx Shearwater						
Newell's Shearwater						
Black-vented Shearwater						
Audubon's Shearwater						
STORKS						
Wood Stork						
FRIGATEBIRDS						
Magnificent Frigatebird						
Great Frigatebird						
BOOBIES AND GANNETS						
Masked Booby						
Brown Booby						
Red-footed Booby						
Northern Gannet						
DARTERS						
Anhinga						
CORMORANTS						
Brandt's Cormorant						
Red-faced Cormorant						
Pelagic Cormorant						
Great Cormorant						

Double-crested Cormorant						
Neotropic Cormorant						
PELICANS						
American White Pelican						
Brown Pelican						
BITTERNS, HERONS, AND ALLIES						
American Bittern						
Least Bittern						
Great Blue Heron						
Great Egret						
Snowy Egret						
Little Blue Heron						
Tricolored Heron						
Reddish Egret						
Cattle Egret						
Green Heron						
Black-crowned Night-Heron						
Yellow-crowned Night-Heron						
IBISES AND SPOONBILLS						
White Ibis						
Glossy Ibis						
White-faced Ibis						
Roseate Spoonbill						
NEW WORLD VULTURES						
California Condor						

Black Vulture						
Turkey Vulture						
OSPREYS						
Osprey						
HAWKS, KITES, EAGLES, AND ALLIES						
White-tailed Kite						
Hook-billed Kite						
Swallow-tailed Kite						
Golden Eagle						
Northern Harrier						
Sharp-shinned Hawk						
Cooper's Hawk						
Northern Goshawk						
Bald Eagle						
Mississippi Kite						
Snail Kite						
Common Black Hawk						
Harris's Hawk						
White-tailed Hawk						
Gray Hawk						
Red-shouldered Hawk						
Broad-winged Hawk						
Hawaiian Hawk						
Short-tailed Hawk						
Swainson's Hawk						

Zone-tailed Hawk						
Red-tailed Hawk						
Rough-legged Hawk						
Ferruginous Hawk						
BARN OWLS						
Barn Owl						
TYPICAL OWLS						
Flammulated Owl						
Whiskered Screech-Owl						
Western Screech-Owl						
Eastern Screech-Owl						
Great Horned Owl						
Snowy Owl						
Northern Hawk Owl						
Northern Pygmy-Owl						
Ferruginous Pygmy-Owl						
Elf Owl						
Burrowing Owl						
Spotted Owl						
Barred Owl						
Great Gray Owl						
Long-eared Owl						
Short-eared Owl						
Boreal Owl						
Northern Saw-whet Owl						

TROGONS						
Elegant Trogon						
KINGFISHERS						
Ringed Kingfisher						
Belted Kingfisher						
Green Kingfisher						
WOODPECKERS AND ALLIES						
Lewis's Woodpecker						
Red-headed Woodpecker						
Acorn Woodpecker						
Gila Woodpecker						
Golden-fronted Woodpecker						
Red-bellied Woodpecker						
Williamson's Sapsucker						
Yellow-bellied Sapsucker						
Red-naped Sapsucker						
Red-breasted Sapsucker						
American Three-toed Woodpecker						
Black-backed Woodpecker						
Downy Woodpecker						
Nuttall's Woodpecker						
Ladder-backed Woodpecker						
Red-cockaded Woodpecker						
Hairy Woodpecker						
White-headed Woodpecker						

Arizona Woodpecker						
Northern Flicker						
Gilded Flicker						
Pileated Woodpecker						
CARACARAS AND FALCONS						
Crested Caracara						
American Kestrel						
Merlin						
Aplomado Falcon						
Gyrfalcon						
Peregrine Falcon						
Prairie Falcon						
PARAKEETS, MACAWS, AND PARROTS						
Monk Parakeet						
Nanday Parakeet						
Green Parakeet						
Mitred Parakeet						
White-winged Parakeet						
Yellow-chevroned Parakeet						
Red-crowned Parrot						
LORIES, LOVEBIRDS, AND AUSTRALASIAN PARROTS						
Rose-ringed Parakeet						
Rosy-faced Lovebird						
BECARDS, TITYRAS, AND ALLIES						
Rose-throated Becard						

TYRANT FLYCATCHERS						
Northern Beardless-Tyrannulet						
Dusky-capped Flycatcher						
Ash-throated Flycatcher						
Great Crested Flycatcher						
Brown-crested Flycatcher						
La Sagra's Flycatcher						
Great Kiskadee						
Sulphur-bellied Flycatcher						
Tropical Kingbird						
Couch's Kingbird						
Cassin's Kingbird						
Thick-billed Kingbird						
Western Kingbird						
Eastern Kingbird						
Gray Kingbird						
Scissor-tailed Flycatcher						
Fork-tailed Flycatcher						
Olive-sided Flycatcher						
Greater Pewee						
Western Wood-Pewee						
Eastern Wood-Pewee						
Yellow-bellied Flycatcher						
Acadian Flycatcher						
Alder Flycatcher						

Willow Flycatcher						
Least Flycatcher						
Hammond's Flycatcher						
Gray Flycatcher						
Dusky Flycatcher						
Pacific-slope Flycatcher						
Cordilleran Flycatcher						
Buff-breasted Flycatcher						
Black Phoebe						
Eastern Phoebe						
Say's Phoebe						
Vermilion Flycatcher						
VIREOS						
Black-capped Vireo						
White-eyed Vireo						
Bell's Vireo						
Gray Vireo						
Hutton's Vireo						
Yellow-throated Vireo						
Cassin's Vireo						
Blue-headed Vireo						
Plumbeous Vireo						
Philadelphia Vireo						
Warbling Vireo						
Red-eyed Vireo						

Yellow-green Vireo						
Black-whiskered Vireo						
MONARCH FLYCATCHERS						
Kauai Elepaio						
Oahu Elepaio						
Hawaii Elepaio						
SHRIKES						
Loggerhead Shrike						
Northern Shrike						
JAYS AND CROWS						
Canada Jay						
Green Jay						
Pinyon Jay						
Steller's Jay						
Blue Jay						
Florida Scrub-Jay						
Island Scrub-Jay						
California Scrub-Jay						
Woodhouse's Scrub-Jay						
Mexican Jay						
Clark's Nutcracker						
Black-billed Magpie						
Yellow-billed Magpie						
American Crow						
Fish Crow						

Chihuahuan Raven						
Common Raven						
VERDINS						
Verdin						
CHICKADEES AND TITMICE						
Carolina Chickadee						
Black-capped Chickadee						
Mountain Chickadee						
Mexican Chickadee						
Chestnut-backed Chickadee						
Boreal Chickadee						
Gray-headed Chickadee						
Bridled Titmouse						
Oak Titmouse						
Juniper Titmouse						
Tufted Titmouse						
Black-crested Titmouse						
LARKS						
Eurasian Skylark						
Horned Lark						
REED WARBLERS						
Millerbird						
SWALLOWS						
Bank Swallow						
Tree Swallow						

Violet-green Swallow						
Northern Rough-winged Swallow						
Purple Martin						
Barn Swallow						
Cliff Swallow						
Cave Swallow						
BUSHTITS						
Bushtit						
BUSH-WARBLERS						
Japanese Bush-Warbler						
LEAF WARBLERS						
Arctic Warbler						
BULBULS						
Red-vented Bulbul						
Red-whiskered Bulbul						
SYLVIID WARBLERS						
Wrentit						
WHITE-EYES						
Warbling White-eye						
LAUGHINGTHRUSHES						
Greater Necklaced Laughingthrush						
Hwamei						
Red-billed Leiothrix						
KINGLETS						
Ruby-crowned Kinglet						

Golden-crowned Kinglet						
WAXWINGS						
Bohemian Waxwing						
Cedar Waxwing						
SILKY-FLYCATCHERS						
Phainopepla						
NUTHATCHES						
Red-breasted Nuthatch						
White-breasted Nuthatch						
Pygmy Nuthatch						
Brown-headed Nuthatch						
CREEPERS						
Brown Creeper						
GNATCATCHERS AND GNATWRENS						
Blue-gray Gnatcatcher						
Black-tailed Gnatcatcher						
California Gnatcatcher						
Black-capped Gnatcatcher						
WRENS						
Rock Wren						
Canyon Wren						
House Wren						
Pacific Wren						
Winter Wren						
Sedge Wren						

Marsh Wren						
Carolina Wren						
Bewick's Wren						
Cactus Wren						
MOCKINGBIRDS AND THRASHERS						
Gray Catbird						
Curve-billed Thrasher						
Brown Thrasher						
Long-billed Thrasher						
Bendire's Thrasher						
California Thrasher						
LeConte's Thrasher						
Crissal Thrasher						
Sage Thrasher						
Northern Mockingbird						
STARLINGS						
European Starling						
Common Myna						
DIPPERS						
American Dipper						
THRUSHES						
Eastern Bluebird						
Western Bluebird						
Mountain Bluebird						
Townsend's Solitaire						

Omao						
Puaiohi						
Veery						
Gray-cheeked Thrush						
Bicknell's Thrush						
Swainson's Thrush						
Hermit Thrush						
Wood Thrush						
Eyebrowed Thrush						
Clay-colored Thrush						
Rufous-backed Robin						
American Robin						
Varied Thrush						
OLD WORLD FLYCATCHERS						
White-rumped Shama						
Bluethroat						
Siberian Rubythroat						
Northern Wheatear						
OLIVE WARBLERS						
Olive Warbler						
WAXBILLS						
African Silverbill						
Java Sparrow						
Scaly-breasted Munia						
Chestnut Munia						

Red Avadavat						
Common Waxbill						
OLD WORLD SPARROWS						
House Sparrow						
Eurasian Tree Sparrow						
WAGTAILS AND PIPITS						
Eastern Yellow Wagtail						
White Wagtail						
Olive-backed Pipit						
Red-throated Pipit						
American Pipit						
Sprague's Pipit						
FRINGILLINE AND CARDUELINE FINCHES AND ALLIES						
Brambling						
Evening Grosbeak						
Akikiki						
Maui Alauahio						
Palila						
Laysan Finch						
Nihoa Finch						
Akohekohe						
Apapane						
Iiwi						
Maui Parrotbill						
Akiapolaau						

Anianiau						
Hawaii Amakihi						
Oahu Amakihi						
Kauai Amakihi						
Hawaii Creeper						
Akekee						
Hawaii Akepa						
Pine Grosbeak						
Gray-crowned Rosy-Finch						
Black Rosy-Finch						
Brown-capped Rosy-Finch						
House Finch						
Purple Finch						
Cassin's Finch						
Yellow-fronted Canary						
Common Redpoll						
Hoary Redpoll						
Red Crossbill						
Cassia Crossbill						
White-winged Crossbill						
Pine Siskin						
Lesser Goldfinch						
Lawrence's Goldfinch						
American Goldfinch						
Island Canary						

LONGSPURS AND SNOW BUNTINGS						
Lapland Longspur						
Chestnut-collared Longspur						
Smith's Longspur						
Thick-billed Longspur						
Snow Bunting						
McKay's Bunting						
EMBERIZIDS						
Rustic Bunting						
TOWHEES AND SPARROWS						
Rufous-winged Sparrow						
Botteri's Sparrow						
Cassin's Sparrow						
Bachman's Sparrow						
Grasshopper Sparrow						
Olive Sparrow						
Five-striped Sparrow						
Black-throated Sparrow						
Lark Sparrow						
Lark Bunting						
Chipping Sparrow						
Clay-colored Sparrow						
Black-chinned Sparrow						
Field Sparrow						
Brewer's Sparrow						

Fox Sparrow						
American Tree Sparrow						
Dark-eyed Junco						
Yellow-eyed Junco						
White-crowned Sparrow						
Golden-crowned Sparrow						
Harris's Sparrow						
White-throated Sparrow						
Sagebrush Sparrow						
Bell's Sparrow						
Vesper Sparrow						
LeConte's Sparrow						
Seaside Sparrow						
Nelson's Sparrow						
Saltmarsh Sparrow						
Baird's Sparrow						
Henslow's Sparrow						
Savannah Sparrow						
Song Sparrow						
Lincoln's Sparrow						
Swamp Sparrow						
Canyon Towhee						
Abert's Towhee						
California Towhee						
Rufous-crowned Sparrow						

Green-tailed Towhee						
Spotted Towhee						
Eastern Towhee						
SPINDALISES						
Western Spindalis						
YELLOW-BREASTED CHATS						
Yellow-breasted Chat						
BLACKBIRDS						
Yellow-headed Blackbird						
Bobolink						
Eastern Meadowlark						
Western Meadowlark						
Orchard Oriole						
Hooded Oriole						
Bullock's Oriole						
Spot-breasted Oriole						
Altamira Oriole						
Audubon's Oriole						
Baltimore Oriole						
Scott's Oriole						
Red-winged Blackbird						
Tricolored Blackbird						
Shiny Cowbird						
Bronzed Cowbird						
Brown-headed Cowbird						

Rusty Blackbird						
Brewer's Blackbird						
Common Grackle						
Boat-tailed Grackle						
Great-tailed Grackle						
WOOD-WARBLERS						
Ovenbird						
Worm-eating Warbler						
Louisiana Waterthrush						
Northern Waterthrush						
Golden-winged Warbler						
Blue-winged Warbler						
Black-and-white Warbler						
Prothonotary Warbler						
Swainson's Warbler						
Tennessee Warbler						
Orange-crowned Warbler						
Colima Warbler						
Lucy's Warbler						
Nashville Warbler						
Virginia's Warbler						
Connecticut Warbler						
MacGillivray's Warbler						
Mourning Warbler						
Kentucky Warbler						

Common Yellowthroat						
Hooded Warbler						
American Redstart						
Kirtland's Warbler						
Cape May Warbler						
Cerulean Warbler						
Northern Parula						
Tropical Parula						
Magnolia Warbler						
Bay-breasted Warbler						
Blackburnian Warbler						
Yellow Warbler						
Chestnut-sided Warbler						
Blackpoll Warbler						
Black-throated Blue Warbler						
Palm Warbler						
Pine Warbler						
Yellow-rumped Warbler						
Yellow-throated Warbler						
Prairie Warbler						
Grace's Warbler						
Black-throated Gray Warbler						
Townsend's Warbler						
Hermit Warbler						
Golden-cheeked Warbler						

Black-throated Green Warbler						
Rufous-capped Warbler						
Canada Warbler						
Wilson's Warbler						
Red-faced Warbler						
Painted Redstart						
CARDINALS, PIRANGA TANAGERS, AND ALLIES						
Hepatic Tanager						
Summer Tanager						
Scarlet Tanager						
Western Tanager						
Flame-colored Tanager						
Northern Cardinal						
Pyrrhuloxia						
Rose-breasted Grosbeak						
Black-headed Grosbeak						
Blue Grosbeak						
Lazuli Bunting						
Indigo Bunting						
Varied Bunting						
Painted Bunting						
Dickcissel						
TANAGERS AND ALLIES						
Red-crested Cardinal						
Yellow-billed Cardinal						

Saffron Finch						
Yellow-faced Grassquit						
Morelet's Seedeater						
FURTHER NOTES						

Because it is widespread, colorful, and attracted to
backyard feeders, the Northern Cardinal is often one of the
first species noted on a North American birder's life list.

CHRONOLOGICAL LIFE LIST

This list allows you to recall the order in which you spotted various species. Each time you see a new one, add its name to the list, perhaps dating each encounter as well. The running tally tracks how many species you've identified, ordered by date—an interesting history to look back on. ■

1	
2	
3	
4	
5	
6	
7	
8	
9	
10	
11	
12	
13	
14	
15	
16	
17	
18	
19	

20	
21	
22	
23	
24	
25	
26	
27	
28	
29	
30	
31	
32	
33	
34	
35	
36	
37	
38	
39	
40	
41	
42	
43	
44	

45	
46	
47	
48	
49	
50	
51	
52	
53	
54	
55	
56	
57	
58	
59	
60	
61	
62	
63	
64	
65	
66	
67	
68	
69	

70	
71	
72	
73	
74	
75	
76	
77	
78	
79	
80	
81	
82	
83	
84	
85	
86	
87	
88	
89	
90	
91	
92	
93	
94	

95	
96	
97	
98	
99	
100	
101	
102	
103	
104	
105	
106	
107	
108	
109	
110	
111	
112	
113	
114	
115	
116	
117	
118	
119	

120	
121	
122	
123	
124	
125	
126	
127	
128	
129	
130	
131	
132	
133	
134	
135	
136	
137	
138	
139	
140	
141	
142	
143	
144	

145	
146	
147	
148	
149	
150	
151	
152	
153	
154	
155	
156	
157	
158	
159	
160	
161	
162	
163	
164	
165	
166	
167	
168	
169	

170	
171	
172	
173	
174	
175	
176	
177	
178	
179	
180	
181	
182	
183	
184	
185	
186	
187	
188	
189	
190	
191	
192	
193	
194	

195	
196	
197	
198	
199	
200	
201	
202	
203	
204	
205	
206	
207	
208	
209	
210	
211	
212	
213	
214	
215	
216	
217	
218	
219	

220	
221	
222	
223	
224	
225	
226	
227	
228	
229	
230	
231	
232	
233	
234	
235	
236	
237	
238	
239	
240	
241	
242	
243	
244	

245	
246	
247	
248	
249	
250	
251	
252	
253	
254	
255	
256	
257	
258	
259	
260	
261	
262	
263	
264	
265	
266	
267	
268	
269	

270	
271	
272	
273	
274	
275	
276	
277	
278	
279	
280	
281	
282	
283	
284	
285	
286	
287	
288	
289	
290	
291	
292	
293	
294	

295	
296	
297	
298	
299	
300	
301	
302	
303	
304	
305	
306	
307	
308	
309	
310	
311	
312	
313	
314	
315	
316	
317	
318	
319	

320	
321	
322	
323	
324	
325	
326	
327	
328	
329	
330	
331	
332	
333	
334	
335	
336	
337	
338	
339	
340	
341	
342	
343	
344	

345	
346	
347	
348	
349	
350	
351	
352	
353	
354	
355	
356	
357	
358	
359	
360	
361	
362	
363	
364	
365	
366	
367	
368	
369	

370	
371	
372	
373	
374	
375	
376	
377	
378	
379	
380	
381	
382	
383	
384	
385	
386	
387	
388	
389	
390	
391	
392	
393	
394	

395	
396	
397	
398	
399	
400	
401	
402	
403	
404	
405	
406	
407	
408	
409	
410	
411	
412	
413	
414	
415	
416	
417	
418	
419	

There's no mistaking an adult Red-headed Woodpecker,
with its white wing patches and crimson hood. Immature
individuals have gray heads. This woodpecker often sallies
from a high perch to catch insects in midair.

INDEX

ABOUT THE WRITER

Noah Strycker, based in Oregon, is an associate editor of *Birding* magazine, a contributor to *National Geographic* magazine, and the author of several popular books about birds, including *National Geographic Birding Basics, The Thing with Feathers,* and *Birding Without Borders.* He also co-authored National Geographic's *Birds of the Photo Ark* and the second edition of *National Geographic Backyard Guide to the Birds of North America.* Upon initial publication of this book, his life list tallied 6,572 species, his North American list included 718 species, and his backyard list was up to 126 species—and counting!

ART CREDITS

Front cover: Golden-winged Warbler by H. Douglas Pratt; back cover: Snow Bunting (top) and American Flamingo (left) by Diane Pierce; American White Pelican (right) by H. Jon Janosik; pages 1, 35, 136: N. John Schmitt; pages 3 (grosbeak), 154, 165, 226: Diane Pierce; pages 3 (cuckoos), 122: H. Douglas Pratt; pages 4, 48 (standing), 81: Jonathan Alderfer; pages 6, 93, 105, 244: Donald L. Malick; page 8: Cynthia J. House; page 22: Kent Pendleton; page 48 (flying): Daniel S. Smith; pages 64, 186: Thomas R. Schultz.

Since 1888, the National Geographic Society has funded more than 14,000 research, conservation, education, and storytelling projects around the world. National Geographic Partners distributes a portion of the funds it receives from your purchase to National Geographic Society to support programs including the conservation of animals and their habitats.

National Geographic Partners, LLC
1145 17th Street NW
Washington, DC 20036-4688 USA

Get closer to National Geographic explorers and photographers, and connect with our global community. Join us today at nationalgeographic.org/joinus

For rights or permissions inquiries, please contact National Geographic Books Subsidiary Rights: bookrights@natgeo.com

Library of Congress Cataloging-in-Publication Data
Names: National Geographic Partners (U.S.), issuing body.
Title: Birder's life list & journal / National Geographic.
Other titles: Birder's life list and journal
Description: Washington, D.C. : National Geographic, [2023] I Includes index. I Summary: "This record book for birders offers blanks for details of a lifetime of species sightings"-- Provided by publisher.
Identifiers: LCCN 2022022483 I ISBN 9781426223167 (paperback)
Subjects: LCSH: Bird watching--North America--Juvenile literature. I Birds--North America--Juvenile literature.
Classification: LCC QL681 .B575 2023 I DDC 598.072/34--dc23/eng/20220622
LC record available at https://lccn.loc.gov/2022022483

ISBN: 978-1-4262-2316-7

Printed in China

22/RRDH/1